Packets with Deadlines

A Framework for Real-Time Wireless Networks

Synthesis Lectures on Communication Networks

Editor
Jean Walrand, *University of California, Berkeley*

Synthesis Lectures on Communication Networks is an ongoing series of 50- to 100-page publications on topics on the design, implementation, and management of communication networks. Each lecture is a self-contained presentation of one topic by a leading expert. The topics range from algorithms to hardware implementations and cover a broad spectrum of issues from security to multiple-access protocols. The series addresses technologies from sensor networks to reconfigurable optical networks.
The series is designed to:

- Provide the best available presentations of important aspects of communication networks.

- Help engineers and advanced students keep up with recent developments in a rapidly evolving technology.

- Facilitate the development of courses in this field.

Packets with Deadlines: A Framework for Real-Time Wireless Networks
I-Hong Hou and P.R. Kumar
2013

Energy-Efficient Scheduling under Delay Constraints for Wireless Networks
Randall Berry, Eytan Modiano, and Murtaza Zafer
2012

NS Simulator for Beginners
Eitan Altman and Tania Jiménez
2012

Network Games: Theory, Models, and Dynamics
Ishai Menache and Asuman Ozdaglar
2011

An Introduction to Models of Online Peer-to-Peer Social Networking
George Kesidis
2010

Stochastic Network Optimization with Application to Communication and Queueing Systems
Michael J. Neely
2010

Scheduling and Congestion Control for Wireless and Processing Networks
Libin Jiang and Jean Walrand
2010

Performance Modeling of Communication Networks with Markov Chains
Jeonghoon Mo
2010

Communication Networks: A Concise Introduction
Jean Walrand and Shyam Parekh
2010

Path Problems in Networks
John S. Baras and George Theodorakopoulos
2010

Performance Modeling, Loss Networks, and Statistical Multiplexing
Ravi R. Mazumdar
2009

Network Simulation
Richard M. Fujimoto, Kalyan S. Perumalla, and George F. Riley
2006

Packets with Deadlines: A Framework for Real-Time Wireless Networks
I-Hong Hou and P.R. Kumar

ISBN: 978-3-031-79256-4   paperback
ISBN: 978-3-031-79257-1   ebook

DOI 10.1007/978-3-031-79257-1

A Publication in the Springer series
SYNTHESIS LECTURES ON COMMUNICATION NETWORKS

Lecture #14
Series Editor: Jean Walrand, *University of California, Berkeley*
Series ISSN
Synthesis Lectures on Communication Networks
Print 1935-4185   Electronic 1935-4193

Packets with Deadlines

A Framework for Real-Time Wireless Networks

I-Hong Hou and P.R. Kumar
Texas A&M University

SYNTHESIS LECTURES ON COMMUNICATION NETWORKS #14

ABSTRACT

With the explosive increase in the number of mobile devices and applications, it is anticipated that wireless traffic will increase exponentially in the coming years. Moreover, future wireless networks all carry a wide variety of flows, such as video streaming, online gaming, and VoIP, which have various quality of service (QoS) requirements. Therefore, a new mechanism that can provide satisfactory performance to the complete variety of all kinds of flows, in a coherent and unified framework, is needed.

In this book, we introduce a framework for real-time wireless networks. This consists of a model that jointly addresses several practical concerns for real-time wireless networks, including per-packet delay bounds, throughput requirements, and heterogeneity of wireless channels. We detail how this framework can be employed to address a wide range of problems, including admission control, packet scheduling, and utility maximization.

KEYWORDS

delay, QoS, real-time communication, admission control, scheduling, utility maximization, wireless

Contents

Preface ix

1 **Introduction** 1
  1.1 Motivation 1
  1.2 Wireless Networks 1
  1.3 Real-Time Systems 4
  1.4 Overview of Book 5

2 **A Study of the Base Case** 7
  2.1 A Basic System Model for Real-Time Wireless Networks 7
  2.2 Feasibility Analysis 10
  2.3 Scheduling Policies 15
  2.4 Proofs of Optimality 17
  2.5 Simulation Results 22

3 **Admission Control** 27
  3.1 An Efficient Algorithm when Packet Generation is Periodic 27
  3.2 Admission Control under Fading Channels 29

4 **Scheduling Policies** 31
  4.1 An Extended System Model 31
  4.2 A Framework for Determining Scheduling Policies 33
  4.3 Scheduling over Unreliable Fading Channels 37
  4.4 Scheduling Policy under Rate Adaptation 39

5 **Utility Maximization without Rate Adaptation** 41
  5.1 Problem Formulation and Decomposition 41
  5.2 A Bidding Procedure between Clients and Access Point 45
  5.3 A Scheduling Policy for the Acess Point 47
    5.3.1 Convergence of the Weighted Transmission Policy 48
    5.3.2 Optimality of the Weighted Transmission Policy 51
  5.4 Simulation Results 52

6   Utility Maximization with Rate Adaptation **55**

6.1   Problem Overview 55

6.2   Examples of Applications 57

    6.2.1   Delay-Constrained Wireless Networks with Rate Adaptation 57

    6.2.2   Mobile Cellular Networks 57

    6.2.3   Dynamic Spectrum Allocation 58

6.3   A Utility Maximization Approach 58

    6.3.1   Convex Programming Formulation 58

    6.3.2   An On-line Scheduling Policy 59

6.4   Incentive Compatible Auction Design 63

    6.4.1   Basic Mechanism and Incentive Compatibility Property 63

    6.4.2   Proof of Optimality 65

    6.4.3   Implementation Issues 66

6.5   Algorithms for Specific Applications 67

    6.5.1   Delay-Constrained Wireless Networks with Rate Adaptation 67

    6.5.2   Mobile Cellular Networks 67

    6.5.3   Dynamic Spectrum Allocation 68

7   Systems with Both Real-Time Flows and Non-Real-Time Flows **71**

7.1   System Overview and Problem Formulation 71

7.2   A Solution Using Dual Decomposition 72

7.3   A Dynamic Algorithm and Its Convergence 73

8   Broadcasting and Network Coding **75**

8.1   System Model 75

8.2   A Framework for Designing Feasibility-Optimal Policies 77

8.3   Scheduling without Network Coding 79

8.4   Broadcasting with XOR Coding 82

8.5   Broadcasting with Linear Coding 84

8.6   Simulation Results 87

A   Lyapunov Analysis and its Application to Queueing Systems **93**

B   Incentive Compatible Auction Design **97**

Bibliography **99**

Authors' Biographies **105**

Preface

Recent years have witnessed the tremendous success and growth of wireless networks. More and more applications are using wireless networks to carry out delay-sensitive tasks. It has become important to address the specific challenges of these applications so as to enhance the quality of service (QoS) and quality of experience (QoE) of end users. This book introduces a new framework for real-time wireless networks for providing service guarantees. Inspired by models that have proven successful for real-time computation, it models real-time wireless networks by postulating a hard delay bound for each packet, a corresponding timely throughput requirement for each flow, and the channel condition of each wireless link. This model is used to develop several mechanisms for serving delay-sensitive applications, including admission control to determine whether it is feasible to satisfy the service requirements of all the clients, packet scheduling for a variety of wireless channels, and utility maximization to determine the optimal resource allocation among the clients. This book also discusses practical challenges in serving systems comprising of both real-time and non-real-time flows, broadcasting real-time flows, and dealing with the selfish behaviors of clients.

In order to address the aforementioned topics, this book exploits a wide range of analytical tools that are important for conducting research on networks and queueing systems. These tools include Blackwell's approachability analysis (Chapter 2), Lyapunov analysis (Chapters 4, 8, and Appendix A), Kelly's decomposition technique (Chapter 5), dual decomposition (Chapter 6), and VCG auctions (Chapter 6 and Appendix B).

I-Hong Hou and P.R. Kumar
April 2013

CHAPTER 1

Introduction

1.1   MOTIVATION

With the advent of broadband wireless transmission, the increasing popularity of mobile devices, and the deployment of wireless sensor networks, wireless networks are increasingly used to serve real-time flows that require strict per-packet delay bounds. Such applications include VoIP, video streaming, real-time surveillance, and networked control. For example, a study by Cisco [1] has predicted that wireless traffic will grow exponentially, and that mobile video will dominate wireless traffic in the near future, accounting for 62% of wireless traffic by the year 2015. Serving real-time flows is also a key component of many cyber-physical systems. In one example of a centralized cyber-physical system, a server may gather surveillance data from wireless sensors, based on which it makes control decisions and disseminates them to actuator units. Both surveillance data and control decisions need to be delivered in a timely manner, otherwise the performance of the system may be degraded. In addition to delay bounds, such applications also require some guarantees on their *timely-throughput*, defined as the throughput of packets that are delivered on time.

Serving real-time flows in wireless networks is particularly challenging since wireless is a shared medium, which raises fundamental issues concerning concurrent transmissions. Also wireless transmissions are subject to shadowing, fading, and interference, and thus usually unreliable. Further, the channel qualities may differ from link to link. A desirable solution for serving real-time flows thus needs to explicitly take into account the shared nature of the medium, and stochastic and heterogeneous behavior of packet losses due to failed transmissions.

In the past, there have been several alternative attempts to model a wireless network consisting of flows that have delay constraints. The difficulty has been that these formulations have by and large led to intractable analytical problems. The end result has been that essentially no significant progress has been made in this area. Therefore, a framework that both captures the practical challenges of real-time wireless networks as well as offers tractable solutions is needed.

1.2   WIRELESS NETWORKS

An important feature of wireless networks is that all transmissions share the same wireless medium. Therefore, transmissions may interfere with each other. A typical way to model the interference is the *protocol model* [2], under which a "collision" occurs and transmissions fail when two interfering wireless links transmit simultaneously. There are two different approaches to avoid collisions. One is a centralized approach where a scheduler chooses a set of non-interfering links to transmit at each time instant. Most cellular network protocols, such as WiMax and LTE, use this approach. Another is a distributed approach where each wireless node chooses whether to transmit or not based on its observation of system history. IEEE 802.11, also known as WiFi, is an example of this distributed approach. In this book, we focuses on the centralized approach. Importantly we show that the system lends itself to such a solution for the case of an access point, both in the case of uplink as well as downlink. This makes implementation relatively easy; as an example, our solution can be realized over the Point Coordination Function mode of IEEE 802.11.

Providing services for flows with delay constraints over wireless links has been gaining extensive research interest. Stockhammer, Jenkac, and Kuhn [3] have studied the minimum initial delay and the minimum required buffer size for video streaming. Their study considers the case where there is only one wireless client in the system. Kang and Zakhor [4] have focused on improving the quality of video streaming by giving priorities to packets according to the content of the video. Li and Schaar [5] have proposed an adaptive algorithm for tuning the MAC retry limit for layered coded video. None of these works provides theoretical understanding on the three important problems for providing services: scheduling algorithms, admission control, and utility maximization.

Scheduling policies for QoS support on error-prone wireless channels have also been of increasingly interest in recent years. Tassiulas and Ephremides [6] have proposed a max weight scheduling policy and proved that it is throughput optimal. Neely [7] has further evaluated this policy and shown that the policy achieves order-optimal average delay. Shakkottai and Stolyar [8] have evaluated various scheduling policies to support a mixture of real-time and non-real-time traffic. Johnsson and Cox [9] have proposed a policy that aims to achieve both low packet delay and high user throughput. Dua and Bambos [10] have focused on the trade-off between user fairness and system performance and designed a policy for this purpose. However, these works do not provide a thorough theoretical study with provable performance guarantees on per-packet delays. Raghunathan et al [11] and Shakkottai and Srikant [12] have developed analytical results on scheduling. However, the goal of their works is to minimize the total number of expired packets over all users, which is inevitably unfair to clients with poor channel qualities. Stolyar and Ramanan [13] aim at offering QoS guarantees on a per-client basis. Their approach offers asymptotic optimality only when the period is large. Kawata et al [14] have focused more on implementation issues and enhancing QoS for the IEEE 802.11 mechanisms. Their simulations have been conducted in a controlled environment where packet losses are infrequent. Fattah and Leung [15] and Cao and Li [16] have provided extensive surveys on scheduling policies for providing QoS.

Compared to scheduling policies, there are fewer analytical studies addressing admission control. Xiao et al [17] and Pong el al [18] have proposed admission control algorithms to guarantee a certain amount of bandwidth for each user but do not take into account any latency constraints. Garg et al [19], Zhai et al [20], and Shin and Schulzrinne [21] have used various metrics to predict system performance statistically but lack a theoretical study. Gao, Cai, and Ngan [22], Niyato and Hossain [23], and Ahmed [24] have surveyed existing admission control algorithms in different types of wireless networks.

There has been much research on utility maximization for both wireline and wireless networks. Kelly [25] and Kelly, Maulloo, and Tan [26] have considered a rate control algorithm to achieve maximum utility in a wireline network. Lin and Shroff [27] have studied the same problem with multi-path routing. For wireless networks, Xiao, Shroff, and Chong [28] have proposed a power-control framework to maximize utility, which is defined as a function of the signal-to-interference ratio. Cao and Li [29] have proposed a bandwidth allocation policy that also considers channel degradation. Bianchi, Campbell, and Liao [30] have studied utility-fair services in wireless networks. However, all the aforementioned works assume that the utility is only determined by the allocated bandwidth. Thus, they do not consider applications that require delay bounds.

Zhang and Du [31] have proposed a cross-layer design for multimedia broadcast. Raghu-nathan et al [11] have proposed scheduling policies for broadcasting delay-constrained flows. This work focuses only on minimizing the total number of expired packets and does not consider the different throughput requirements on different flows for each client. Gopala and El Gamal [32] have studied the tradeoff between throughput and delay of broadcasting. They have only studied the scaling laws for average delay, and thus their results are not applicable to scenarios where strict per-packet delay bounds are required. Zhou and Ying [33] have studied the asymptotic capacity of delay-constrained broadcast in mobile ad-hoc networks.

In recent years, network coding has emerged as a powerful technique to improve the capacity of wireless networks. Chaporkar and Proutiere [34] have proposed an adaptive network coding policy to improve throughputs of multi-hop unicast flows. Ghaderi, Towsley, and Kurose [35] have quantified the reliability gain of network coding for broadcasting in unreliable wireless environments. Nguyen et al [36] have compared the throughputs of broadcast flows in systems employing network coding with those without network coding. Lucani, Medard, and Stojanovic [37] have analyzed the computational overhead of using different network coding schemes. Kozat [38] has studied the throughput capacity when erasure codes are employed. These works focus on throughputs and do not consider delays. Yeow, Hoang, and Tham [39] have focused on minimizing delay for broadcast flows by using network coding. Eryilmaz, Ozdaglar, and Medard [40] have studied the gain in delay performance resulting from network coding. Ying, Yang, and Srikant [41] have demonstrated that coding achieves the optimal delay-throughput tradeoff in mobile ad-hoc networks. These works only consider the performance of average delays and do not address strict per-packet delay bounds. Li, Wang, and Lin [42] have studied a special case where a basestation broadcasts delay-constrained flows to two clients. They demonstrate that using opportunistic network coding achieves the maximum asymptotic throughput for this special case. Both Pu el al [43] and Gangammanavar and Eryilmaz [44] have studied optimal coding strategies for broadcasting delay-constrained flows. Their works require the basestation to obtain feedback information from clients frequently, and thus may not be scalable.

1.3 REAL-TIME SYSTEMS

There is a rich literature in the area of real-time systems. Loosely speaking, a real-time system is required to complete its work and deliver its services on a timely basis [45]. In their seminal paper [46], Liu and Layland have studied the classic periodic hard real-time model. In this model, there are a number of *tasks* in the system. Task n generates *jobs* periodically with period τ_n. That is, task n generates jobs at time $0, \tau_n, 2\tau_n, \ldots$. Each job of task n needs to be finished before the next job of n is generated. Thus, the job that is generated at time $k\tau_n$ needs to be finished before time $(k+1)\tau_n$. Further, it is assumed that all jobs of task n require a total amount of time c_n to be finished. Liu and Layland establish two important results for this model. First, they show that there exists a feasible schedule that finishes all jobs on time, as long as $\sum_n \frac{c_n}{\tau_n} \leq 1$. Second, they show that the Earliest Deadline First (EDF) scheduling policy finishes all jobs on time, as long as that is feasible.

The hard real-time model has been extended to consider scenarios where a deadline miss only results in system performance degradation, instead of a fatal error. The imprecise computation model [47, 48] has been proposed to handle applications in which partially completed jobs are useful. In this model, all jobs consist of two parts: a mandatory part and an optional part. The mandatory part needs to be completed before its deadline, or else the system suffers from a timing fault. On the other hand, the optional part is used to further enhance performance by either reducing errors or increasing rewards. The relations between the errors, or rewards, and the time spent on the optional parts, are described through error functions or reward functions. Chung, Liu, and Lin [47] have proposed scheduling policies that aim to minimize the total average error in the system for this model. Their result is optimal only when the error functions are linear and tasks generate jobs with the same period. Shih and Liu [49] have proposed policies that minimize the maximum error among all tasks in the system when error functions are linear. Feiler and Walker [50] have used feedback to opportunistically schedule the optional parts when the lengths of the mandatory parts may be time-varying. Mejia-Alvarez, Melhem, and Mosse [51] have studied the problem of maximizing total rewards in the system when job generations are dynamic. Chen et al [52] have proposed scheduling policies that defer optional parts so as to provide more timely response for mandatory parts. Zu and Chang [53] have studied the scheduling problem when optional parts are hierarchical. Aydin et al [54] have proposed an off-line scheduling policy that maximizes total rewards when the reward functions are increasing and concave. Most of these works only concern themselves with the maximization of the total reward in a system. Amirijoo, Hansson, and Son [55] have considered the tradeoff between data errors and transaction errors in a real-time database. The Increasing Reward with Increasing Service (IRIS) models can be thought of as special cases of the imprecise computation models where the lengths of mandatory parts are zero. Scheduling policies aimed at maximizing total rewards have been studied for such models [56, 57].

1.4 OVERVIEW OF BOOK

In this book, we introduce a useful and tractable framework for the modeling, analysis, and design of real-time wireless networks. This formulation is also generalizable in several directions to handle various additional features, while still providing tractable solutions, and, in some cases, somewhat surprising answers. This framework is built on top of an analytical model that jointly considers the three important aforementioned challenges:

1. a strict deadline for each packet,

2. the timely-throughput requirement specified by each client or application,

3. the unreliable and heterogeneous nature of wireless transmissions.

An important feature is that this model is suitable for characterizing the needs of a wide range of applications, and allows each application to specify its individual requirements. Thus the contracts that result from this framework are on the one hand supportable by protocols, and on the other relevant ant utilizable by application designers.

Using this framework, we provide several important solutions for serving applications requiring both delay guarantees and timely-throughput requirements. We provide a simple and sharp characterization of when the demands of all the clients in a system are feasible under the joint limitations of their channel reliabilities, and obtain a polynomial-time algorithm for admission control. Further, we address the problem of packet scheduling. We establish an on-line scheduling policy that is feasibility optimal in the sense that it fulfills the demands of any set of clients as long as the demands of the set of clients are feasible [58, 59].

This framework can be further generalized in various directions of interest. We show that it can be extended to address more complicated and realistic channel models, including unreliable fading channels and the usage of rate adaptation. We introduce a general method for designing scheduling policies for different channel models, and show that this method results in tractable policies for various cases [60].

We also extend the framework to address scenarios where the timely-throughput requirements of clients are elastic. This can be formulated as a utility maximization problem. A bidding game, in which both clients and servers follow simple on-line strategies, then achieves the maximum total utility in the system [61].

The problem of utility maximization can also be addressed under this framework when rate adaptation is used. We further study the problem when behaviors of clients can be selfish and strategic, and design an "incentive compatible auction," under which strategic clients gain nothing by hiding their private preferences. We show that these results can also be applied to other wireless applications, including dynamic spectrum allocation and mobile wireless cellular networks [62].

Another extension is to consider scenarios where both real-time flows and non-real-time flows are present. We cover a model analyzed in [63] that aims to maximize the total utility of non-real-time flows while providing timely-throughput requirements for real-time ones. This problem can be optimally solved by using a dual decomposition technique.

We also generalize this framework to "broadcast" flows, i.e., flows where a single transmission has multiple recipients, that require delay guarantees. We show that the resulting model can also accommodate the optional usage of various network coding mechanisms. We then design scheduling policies for several different network coding mechanisms [64].

CHAPTER 2

A Study of the Base Case

2.1   A BASIC SYSTEM MODEL FOR REAL-TIME WIRELESS NETWORKS

We start by describing a basic model that can incorporate delay bounds and delivery ratio requirements for real-time flows, as well as the unreliable and heterogeneous nature of wireless channels. We will show that the model can address both uplink and downlink traffic.

Consider a wireless system with N wireless clients, $\{1, 2, \ldots, N\}$, and one access point (AP). Each wireless client generates a real-time flow. Time is slotted with slots denoted by $t \in \{0, 1, 2, \ldots\}$. The length of a time slot is chosen so that it is large enough to accommodate the time needed for transmitting a packet and all overheads, which will be further explained in the next paragraph. The AP is in charge of scheduling transmissions in the time slots. There can be at most one such transmission in a time slot since two concurrent transmissions result in a collision.

We consider both downlink, i.e., from the AP to a client, and uplink, i.e., from a client to the AP. When the AP schedules a downlink slot for client n, it broadcasts a data packet intended for client n. If client n receives the data packet successfully, it broadcasts an ACK intended for the AP, to indicate that it has received the packet. The AP considers the downlink packet to have been successfully delivered if it receives the ACK from the client. Thus, the total time needed for a downlink packet delivery includes the time needed for sending a data packet, as well as that for sending back an ACK. Figure 2.1 illustrates such a downlink packet transaction.

On the other hand, when the AP schedules a uplink slot for client n, it first transmits a POLL packet intended for client n. This POLL packet indicates that client n can transmit in that time slot. If client n receives the POLL packet correctly, it transmits a data packet intended for the AP. The uplink packet is considered to have been successfully delivered if the AP receives a data packet from client n successfully. Figure 2.2 illustrates such an uplink transaction. Note that there are no ACKs involved in uplink transactions. Since we consider the scenario where the AP is in charge of scheduling all transmissions, only the AP needs to know whether a transmission is successful or not. Hence, the AP does not need to send an ACK back to the client in the case of uplink. We also note that the time needed for an uplink transaction is similar to the time needed for a downlink transaction, since both ACK and POLL are small packets that need a much smaller time to transmit than the data packet.

In the sequel, by the word "transmission" we will mean the sequence of two transmissions involved in a slot.

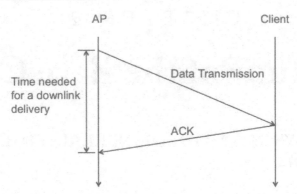

Figure 2.1: An illustration of the transmissions involved in a downlink packet delivery

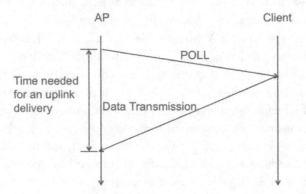

Figure 2.2: An illustration of the transmissions involved in an uplink packet delivery

Time slots are further grouped into *intervals*, each consisting of T time slots. The k-th interval consists of the T consecutive time slots in $[kT, (k + 1)T)$. At the beginning of each interval, that is, at times $1, T + 1, 2T + 1, \ldots$, each real-time flow generates one packet. We assume that the packets in all the real-time flows need to be delivered within a delay bound of T time slots if they are to be useful. In other words, packets that are generated at the beginning of an interval are only useful if they are delivered no later than the end of that interval. If a packet is not delivered before its delay bound, the packet is considered to have "expired" and is dropped from the system. By dropping expired packets, it is guaranteed that the delay of every delivered packet is at most T time slots.

We allow unreliable and heterogeneous wireless links. When the AP makes a transmission for client n, the transmission is successful with probability p_n. (Keeping in mind that we have clubbed together two individual transmissions, i.e., DATA and ACK in the case of downlink, or POLL and DATA in the case of uplink, the quantity p_n represents the probability that both transmissions involved in a slot are successfully received by their intended recipients.) The values of p_n may differ from client to client, as the channel reliabilities for different clients may be different. As explained previously, the AP has instant feedback information on whether a transmission, either an uplink one or a downlink one, is successful. If a transmission fails, the AP may retransmit the same packet in the next time slot if it chooses to, as long as the packet has not expired.

Figure 2.3 illustrates an example of the system where we have one AP and three clients, and the length of an interval is $T = 5$ time slots. Suppose that in the first interval the AP first schedules client 3 for transmission. Due to the unreliability of wireless transmissions, suppose that the transmission fails, and that the AP schedules client 3 for transmission again in the second time slot, which turns out to be successful. Next, suppose that the AP schedules client 2 for transmission in the third time slot, and the transmission succeeds. Then, the AP schedules client 1 on the fourth and fifth time slots of the interval, and both transmissions fail. At this point, the first interval ends, but the packet of client 1 has not been delivered. Hence, the packet of client 1 expires and is removed from the system.

In the second interval, the AP may schedule client 2 in the first time slot, client 3 in the second time slot, and client 1 in the third time slot. This time, all three transmissions are successful, and all packets are delivered in the first three time slots. For the remaining two time slots in the interval, there are no packets to be transmitted. Therefore, the fourth and fifth time slots in the second interval are forced to be idle. This enforced idling is an important feature that we will consider in detail in the sequel.

In this example, we see two scenarios that can transpire in an interval. First, some packets cannot be delivered before the end of the interval. These packets expire, and are removed from the system. Also, if all packets are delivered before the end of the interval, and there are some time slots still left in the interval, then the channel is forced to be idle for the remainder of the interval.

We measure the performance of a client by its *timely-throughput*, which is defined as the long-term average number of successfully delivered packets for the client per interval. Specifically, if $e_n(k)$ is the indicator function that there is a successful packet delivery for client n in the kth interval, then the timely-throughput of client n is $\liminf_{K \to \infty} \frac{\sum_{k=1}^{K} e_n(k)}{K}$. In the example in Fig. 2.3, there is one packet delivered for client 1 in the two intervals. Hence, the timely-throughput of client 1 is $\frac{1}{2}$. (Of course in the example we have only illustrated the case of finite $K = 2$ without taking the limit as $K \to \infty$ as we properly should in order to calculate the timely-throughput.) On the other hand, there are two packets delivered for each of clients 2 and 3. Hence, the timely-throughputs of both clients 2 and 3 are 1.

We assume that each client has an inelastic *timely-throughput requirement*. We denote the timely-throughput requirement of client n by q_n. In other words, each client n requires that $\liminf_{K \to \infty} \frac{\sum_{k=1}^{K} e_n(k)}{K} \geq q_n$, almost surely.

In the following chapters, we will discuss how to characterize whether it is *feasible* to *fulfill* the requirements of all the clients in the system, and how to design a *feasibility optimal* policy that

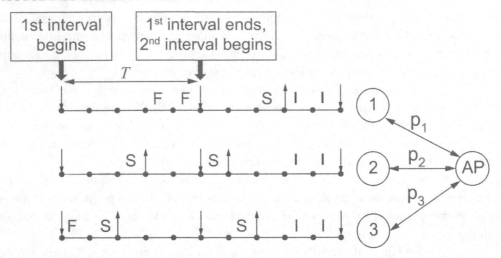

Figure 2.3: An example that illustrates the system model. The right half of the figure shows the topology that includes one AP and three clients. The left half of the figure shows the timeline of each client. We use a downward arrow to indicate a packet arrival, an upward arrow to indicate a packet delivery, 'F' to indicate a failed transmission, and 'S' to indicate a successful transmission.

fulfills the requirements of a set of clients as long as the set is *strictly feasible*. These terms are formally defined as follows.

Definition 2.1   A system is said to be *fulfilled* by some scheduling policy η, if, under η, the timely-throughput provided to each client n is at least q_n.

Definition 2.2   A system is *feasible* if there exists some scheduling policy that fulfills it.

Definition 2.3   A scheduling policy is *feasibility optimal* if it fulfills every feasible system.

2.2   FEASIBILITY ANALYSIS

In this section, we derive a necessary condition for a set of clients to be feasible. We first consider the scenario where there is exactly one client.

Example 2.4   Consider a system with only one client. That is, $N = 1$. Since there is only one client in the system, the feasibility optimal policy is obviously one that schedules client 1 whenever there is a packet to be transmitted. To compute the timely-throughput of client 1, we notice that client 1 has a packet delivery in an interval unless all T transmissions in the interval fails, which occurs with probability $(1 - p_1)^T$. Thus, the timely-throughput of client 1 is $1 - (1 - p_1)^T$, and the system is feasible if and only if $q_1 \leq 1 - (1 - p_1)^T$.

However, the same analysis is not sufficient for systems with more than one client. Therefore, we need a different approach for analyzing systems with multiple clients.

We begin by noting that the more often the server schedules a client, the higher the timely-throughput that the client gets. More formally, we have the following lemma:

Lemma 2.5   *The long-term average timely-throughput of a client n is at least q_n packets per interval if and only if the AP, on average, schedules that client $w_n := w_n(q_n) = \frac{q_n}{p_n}$ times per interval.*

The proof of Lemma 2.5 is based on the law of large numbers for martingales [65]. We first introduce some basic concepts of martingales [66].

Definition 2.6   Let $X_1, X_2, \ldots, X_t, \ldots$ be a series of random variables, where the distribution of X_{t+1} may depend on $X_1, X_2, \ldots, X_t$. If

$$E[X_{t+1}|X_1, X_2, \ldots, X_t] = X_t, \tag{2.1}$$

then we call $\{X_t\}$ a *martingale*. If

$$E[X_{t+1}|X_1, X_2, \ldots, X_t] \geq X_t, \tag{2.2}$$

then we call $\{X_t\}$ a *submartingale*.

Theorem 2.7   (Sub)martingale Convergence Theorem   *If $\{X_t\}$ is a (sub)-martingale satisfying*

$$\sup_t E[X_t^+] < \infty, \tag{2.3}$$

where $X_t^+ := \max\{0, X_t\}$, then there exists a random variable X_∞ with $E[X_\infty] < \infty$ such that

$$\lim_{t \to \infty} X_t \to X_\infty, \tag{2.4}$$

almost surely.

Let $\Delta X_t := X_t - X_{t-1}$. If $\{X_t\}$ is a martingale, we have

$$E[\Delta X_{t+1}|X_1, \Delta X_2, \Delta X_3, \ldots, \Delta X_t] = E[X_{t+1} - X_t|X_1, X_2, \ldots, X_t] = 0. \tag{2.5}$$

We also have:

Theorem 2.8   Law of large numbers for martingales   *Let $\{X_t\}$ be a martingale. If $b_1 < b_2 < \cdots \to \infty$ and*

$$\sum_{t=1}^{\infty} E[(\Delta X_t)^2]/b_t^2 < \infty, \tag{2.6}$$

then

$$\lim_{t \to \infty} X_t/b_t = 0, \tag{2.7}$$

almost surely. In particular, we can set $b_t \equiv t$ in (2.6), and then we have

$$\lim_{t \to \infty} X_t/t = 0, \tag{2.8}$$

if

$$\sum_{t=1}^{\infty} E[\Delta X_t^2]/t^2 < \infty, \tag{2.9}$$

We are now ready to prove Lemma 2.5.

Proof of Lemma 2.5. Define:

$$u_n(t) = \begin{cases} 1, & \text{if client } n \text{ makes a transmission at time } t, \\ 0, & \text{otherwise,} \end{cases}$$

and

$$d_n(t) = \begin{cases} 1, & \text{if client } n \text{ delivers a packet at time } t, \\ 0, & \text{otherwise.} \end{cases}$$

Let $\mathfrak{F}_t$ be the σ-algebra generated by $\{(u_n(k), d_n(k-1)), \text{ for } 1 \le k \le t \text{ and } 1 \le n \le N\}$. (We set $d_n(0) = 0$ for all n.)

Then $E[d_n(t)|\mathfrak{F}_t] = p_n u_n(t)$. Letting $\Delta X_t := d_n(t) - p_n u_n(t)$ and $X_t := \sum_{s=1}^{t} \Delta X_s$, we then have

$$E[X_{t+1}|X_1, X_2, \ldots, X_t] = E[\Delta X_{t+1} + X_t|X_1, X_2, \ldots, X_t] = X_t, \tag{2.10}$$

and so $\{X_t\}$ is a martingale. Moreover, as $|d_n(t) - p_n u_n(t)| < 1$, we have

$$\sum_{t=1}^{\infty} E[\Delta X_t^2]/t^2 \le \sum_{t=1}^{\infty} 1/t^2 < \infty. \tag{2.11}$$

Therefore, by the law of large numbers for martingales, we have

$$\lim_{\mathfrak{T} \to \infty} \frac{1}{\mathfrak{T}} \sum_{t=1}^{\mathfrak{T}} [d_n(t) - p_n u_n(t)] = 0, \text{ a.s.} \tag{2.12}$$

Therefore, from 2.12, the timely-throughput of client n is at least q_n, i.e.,

$$\liminf_{\mathfrak{T} \to \infty} \frac{T}{\mathfrak{T}} \sum_{t=1}^{\mathfrak{T}} d_n(t) \ge q_n, \text{ a.s.} \tag{2.13}$$

if and only if $\liminf_{\mathfrak{T} \to \infty} \frac{T}{\mathfrak{T}} \sum_{t=1}^{\mathfrak{T}} u_n(t) \ge \frac{q_n}{p_n}$. $\qquad\square$

We will hereafter refer to w_n as the *workload for client n*. Thus, a set of clients is fulfilled if and only if the average numbers of transmissions per interval for packets for each client is higher than its workload.

Since the delay bound for all clients is T time slots, and since the AP can make at most one transmission in each time slot, the following necessary condition is obtained:

Lemma 2.9 *A set of N clients is feasible only if $\sum_{n=1}^{N} w_n \leq T$.*

This necessary condition turns out, however, to be not sufficient. Consider a system with one client. Example 2.4 has shown that the system is feasible if and only if $q_1 \leq 1 - (1 - p_1)^T$. On the other hand, the condition in Lemma 2.9 only requires that $q_1 \leq p_1 T$. Since we have $1 - (1 - p_1)^T < T p_1$, for all $T \geq 2$ and $0 < p_1 \leq 1$, the condition in Lemma 2.9 is clearly not sufficient.

To see why the condition in Lemma 2.9 is not sufficient, we can refer to the example in Fig. 2.3. In the second interval, all packets are delivered in the first three time slots. Since undelivered packets expire and are discarded at the end of each interval, the AP can only transmit packets that are generated in the current interval. As a result, the system is forced to be idle for the last two time slots in the interval. Therefore, to determine whether a system is feasible, we need to take these unavoidable idle time slots into account.

It is easy to see that the number of idle time slots is the same for the following set of policies:

Definition 2.10 A scheduling policy is said to be *work conserving* if the AP never idles whenever there is any undelivered packet.

Let γ_n be the random variable denoting the number of transmissions the AP needs to make for a packet for client n before it is successfully delivered. γ_n has the geometric distribution with parameter p_n, that is, $Prob\{\gamma_n = t\} = p_n(1 - p_n)^{t-1}$ for all positive integers t.

Lemma 2.11 *The probability distribution of the amount of idle time slots in an interval is the same for all work conserving scheduling policies, and is equal to $E(T - \sum_n \gamma_n)^+$, where $x^+ := max\{0, x\}$.*

Proof. Let L_η be the random variable indicating the number of idle time slots in such an interval under scheduling policy η. We have:

$$L_\eta = (T - \sum_n \gamma_n)^+, \tag{2.14}$$

for all work conserving policies. Thus, the probability distribution of L_η is the same for all work conserving policies. $\square$

Let us I to be the long-term average number of idle time slots in an interval under any work conserving policy.

The following observation shows that we can always construct a work conserving policy, from any policy, by modifying it so that it performs at least as well as the original policy.

Lemma 2.12   *Let η be a scheduling policy that fulfills some sets of clients. Then there exists a work conserving policy η' that fulfills the same set of clients.*

Proof. The policy η can be modified into a work conserving one by attempting any undelivered packet whenever η idles. This modification cannot reduce the number of deliveries for any client and thus would fulfill any set of clients that η fulfills.   □

Based on this lemma, we can therefore limit our discussion to work conserving policies throughout the rest of this chapter. Since, on average, the AP can only make $T - I$ transmissions in an interval, we obtain that a system is feasible only if $\sum_{n=1}^{N} w_n \leq T - I$.

It turns out that this condition too is only necessary and not sufficient for feasibility. It can be further refined. Observe that, if we remove some clients from a feasible system, resulting in a system that only consists of a subset of clients of the original one, the resulting system must also be feasible. Thus, by letting I_S denote the long-term average number of idle time slots in an interval when only a subset $S \subseteq \{1, 2, \ldots, N\}$ of clients is present, we obtain an even more stringent necessary condition.

Lemma 2.13   *A system is feasible only if $\sum_{n \in S} w_n \leq T - I_S$, for all $S \subseteq \{1, 2, \ldots, N\}$.* □

We note that the conditions $\sum_{n=1}^{N} w_n \leq T - I$, and $\sum_{n \in S} w_n \leq T - I_S$, for all $S \subseteq \{1, 2, \ldots, N\}$, are not equivalent. The following example demonstrates that the condition in Lemma 2.13 is indeed a stronger one.

Example 2.14   Consider a system with two clients and interval length $T = 3$. Suppose that the reliabilities for both clients are $p_1 = p_2 = 0.5$. Suppose also that client 1 requires a timely-throughput of $q_1 = 0.876$, while the timely-throughput requirement of client 2 is $q_2 = 0.45$.

Then, we have:

$$w_1 = \frac{q_1}{p_1} = 1.76,$$
$$w_2 = \frac{q_2}{p_2} = 0.9,$$
$$I_{\{1\}} = I_{\{2\}} = 2p_1 + (1 - p_1)p_1 = 1.25,$$
$$I \equiv I_{\{1,2\}} = p_1 p_2 = 0.25.$$

If we evaluate the condition for the set of all clients $\{1, 2\}$, we have $w_1 + w_2 = 2.66 < 2.75 = T - I_{\{1,2\}}$. However, if we evaluate the condition for the subset of $S = \{1\}$, we find $w_1 = 1.76 >$

$1.75 = T - I_{\{1\}}$. This indicates that the set of clients is not feasible. Thus, merely evaluating the condition for the set of all clients is not enough. □

Surprisingly, we will show that the condition in Lemma 2.13 is actually both necessary and sufficient.

2.3 SCHEDULING POLICIES

In this section, we introduce two scheduling policies. We will prove that they are both feasibility optimal, that is, they fulfill every feasible system. Both policies are what we call as *largest debt first scheduling policies*. When employing a largest debt first scheduling policy, the AP calculates the *debt* it owes to each client at the beginning of each interval. The AP then prioritizes all clients according to the debts owed to them, such that clients with larger debts get higher priority. In each time slot of the interval, the AP transmits the packet for the client with the largest debt from among those whose packets are not delivered yet.

Figure 2.4 shows an example of scheduling using the largest debt first scheduling policy. In this example, we assume that, at the beginning of some interval, client 1 has the largest debt, client 3 has a medium debt, and client 2 has the smallest debt. The AP first transmits the packet for client 1, and keeps retransmitting the packet in case the previous transmission fails. The packet for client 3 is only transmitted after the packet for client 1 is successfully delivered. Finally, the packet for client 2 is only transmitted after both packets for client 1 and client 3 are successfully delivered.

The two scheduling policies differ in their definitions of debt. The first kind of debt, the *time-based debt*, is derived from the concept of load w_n defined in Lemma 2.5.

Definition 2.15 Let $u_n(k)$ be the number of time slots that the AP spends in transmitting packet for client n in the kth interval. The *time-based debt* of client n at the beginning of the $(k + 1)^{th}$ interval is defined as $r_n^{(1)}(k + 1) := kw_n - \sum_{j=1}^{k} u_n(j)$. The largest debt first policy that employs the time-based debt is called the *largest time-based debt first* scheduling policy.

In the above definition, kw_n is the number of time slots that the AP *should have* spent transmitting packets for n in the first k intervals, by Lemma 2.5, while $\sum_{j=1}^{k} u_n(j)$ is the number of time slots that the AP *actually* spent transmitting for n. The definition of $r_n^{(1)}(k + 1)$ is then the deficiency of service for client n in terms of the number of scheduled time slots, and hence the name "time-based debt."

The second kind of debt, the *weighted-delivery debt*, is derived directly from the timely-throughput requirement q_n of a client.

Definition 2.16 Let $d_n(k)$ be the indicator function of the event that the AP delivers a packet for client n in the kth interval. The *weighted-delivery debt* of client n at the beginning of the $(k + 1)^{th}$ interval is defined as $r_n^{(2)}(k + 1) := \frac{kq_n - \sum_{j=1}^{k} d_n(j)}{p_n}$. The largest debt first policy that prioritizes

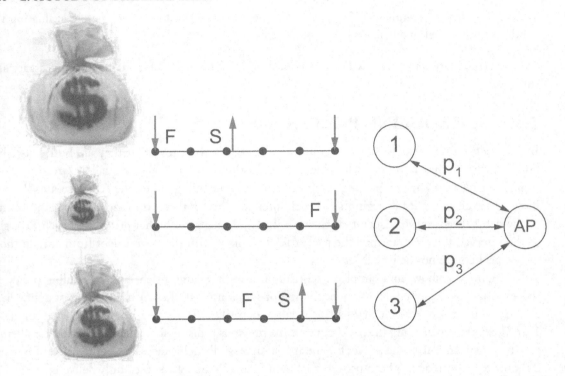

Figure 2.4: An example illustrating the largest debt first scheduling policy. The size of the money bag indicates the amount of debt that the AP owes to the client at the beginning of this interval. We use 'F' to denote a failed transmission, and 'S' to denote a successful one.

according to the weighted-delivery debt is called the *largest weighted-delivery debt first* scheduling policy.

As client n requires a timely-throughput of q_n packets per interval, kq_n is the number of packet deliveries required by n in the first k intervals. Hence, $kq_n - \sum_{j=1}^{k} d_n(j)$ is the deficiency of packet deliveries for client n. We scale this deficiency by $\frac{1}{p_n}$ to simplify the proof of Theorem 2.21.

Next, we provide two examples that illustrate how the two largest debt first scheduling policies work.

Example 2.17   Consider a system with three clients and $T = 10$. Assume that we have $p_1 = p_2 = p_3 = 0.4$ and $q_1 = q_2 = q_3 = 0.8$. Thus, we also have $w_1 = w_2 = w_3 = 0.8/0.4 = 2$. We first consider the case when the AP employs the largest time-based debt first policy. Assume that, at the beginning of an interval, the time-based debts of the three clients are $r_1^{(1)} = 5, r_2^{(1)} = 1$, and $r_3^{(1)} = 3$. Since client 1 has the largest debt, the AP will keep scheduling client 1 until its packet is delivered. Assume that the packet of client 1 is delivered after 5 transmissions, that is, the first four transmissions fail and the fifth transmission succeeds. Next, the AP schedules client 3 at the sixth time slot of the interval, as client 3 has the second largest debt among the three clients. Assume that the transmission at the sixth time slot is successful. Finally, starting from the seventh time slot of the interval, the AP schedules client 2. Assume that the four consecutive transmissions, that is, transmissions at the seventh time slot through the 10th time slot, for client 2 fail. The packet for client 2 has reached its delay bound, and is hence dropped from the system.

In sum, during this interval, client 1 is scheduled four times and delivers a packet, client 2 is scheduled four times without delivering any packet, and client 3 is scheduled once and delivers a packet. At the beginning of the next interval, the time-based debts of the three clients become $r_1^{(1)} = 5 + w_1 - 4 = 5 + 2 - 4 = 3, r_2^{(1)} = 1 + 2 - 4 = -1$, and $r_3^{(1)} = 3 + 2 - 1 = 4$. Hence, in the next interval, the AP will schedule client 3 first, followed by client 1, and then followed by client 2.

Next, we consider the case when the AP employs the largest weighted delivery debt first policy. Assume that, at the beginning of an interval, the time-based debts of the three clients are $r_1^{(2)} = 5$, $r_2^{(2)} = 1$, and $r_3^{(2)} = 3$. Hence, the AP will schedule the clients as in the previous paragraphs, and we assume that events in the interval are exactly the same as those in the previous paragraphs, that is, client 1 is scheduled four times and delivers a packet in the interval, etc. At the beginning of the next interval, we have $r_1^{(2)} = 5 + (q_1 - 1)/p_1 = 5 + (0.8 - 1)/0.4 = 4.5, r_2^{(2)} = 1 + 0.8/0.4 = 3$, and $r_3^{(2)} = 3 + (0.8 - 1)/0.4 = 2.5$. Thus, in the next interval, the AP will first schedule client 1, followed by client 2, and then followed by client 3. □

We note that, in the language of control theory, the largest time-based debt first scheduling policy is more of an "open-loop" policy while the largest weighted-delivery debt first scheduling policy is more of a "closed-loop" policy. The reason is that the delivery debt based policy more fully exploits the information about how many of the transmissions were successful.

2.4   PROOFS OF OPTIMALITY

In this section, we prove that both policies above are feasibility optimal for fairly general traffic patterns. Our proof is based on Blackwell's approachability theorem [67]. We first describe the content of this theorem.

Consider a single player game with a *vector* payoff in each round determined by a probability distribution on the Euclidean N-dimensional space, which in turn depends on the action taken by the player in that round. Suppose that, under some policy, the player takes action $a(i)$ and obtains a payoff $v(i)$, which is an N-dimensional vector, in each round i. Blackwell studied the long-term average payoff the player obtains, that is, $\lim_{j\to\infty} \sum_{i=1}^{j} v(a(i))/j$, and introduced the concept of *approachability*:

Definition 2.18   Let $A \subseteq \mathbb{R}^N$ be any set in the N-dimensional space. Consider a policy η, which incurs payoffs:

$$v(a(1)), v(a(2)), \dots.$$

Let δ_j be the Euclidean distance between the point $\sum_{i=1}^{j} v(a(i))/j$ and A. We say A is *approachable* under policy η, if for every $\varepsilon > 0$ there is a j_0 such that,

$$Prob\{\delta_j \geq \varepsilon \text{ for some } j \geq j_0\} \leq \varepsilon.$$

Blackwell derived a sufficient condition for approachability:

Theorem 2.19   *Let $A \subseteq \mathbb{R}^N$ be any closed set in N-dimensional space. Let η be a policy whose action depends solely on the average payoff to date, $x_j := \sum_{i=1}^{j-1} v(a(i))/j$. Thus, we can express $a(j)$ as a function $a'(x_j)$ depending only on x_j. Then A is approachable under η if η satisfies the following property:*

If $x_j \notin A$, let y be the closest point in A to x_j, and let H be the hyperplane passing through y and perpendicular to the line segment $x_j y$.

Then, under η, H separates x_j and the expected payoff $E[v(a'(x_j))]$ of round j.

Figure 2.5 shows an example that illustrates the relations between x_j, y, H, and $E[v(a'(x_j))]$.

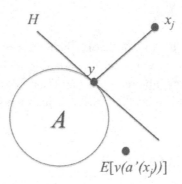

Figure 2.5: An example to illustrate the quantities used in Theorem 2.19

Utilizing this, we now prove that both largest debt first policies are feasibility optimal. Since a feasible set of clients must satisfy the necessary condition in Lemma 2.13, we only need to prove that the two policies fulfill every set of clients that satisfy the necessary condition.

Theorem 2.20   *The largest time-based debt first policy is feasibility optimal.*

Proof. We first translate the model into a single player game. A round in this game corresponds to an interval containing T time slots in the model. The player is the AP. The action the player can take is to choose the priority order of clients for that interval, with the interpretation that an undelivered packet for a client is transmitted in an interval only after all packets from clients with higher priorities are delivered in that interval. The payoff the player obtains in a round is the net change of the time-based debt owed to each client, which is thus an N-dimensional vector. To be more precise, the payoff the player obtains in a round is $v = [v_1, v_2, \ldots, v_N]$, where v_n equals w_n minus the number of times the AP transmits the packet for client n during that interval.

By Lemma 2.5, the demand of a client n is met if the AP transmits its packets at least $w_n = q_n/p_n$ times per interval on average, i.e., equivalently, $\liminf_{K \to \infty} \frac{\sum_{k=1}^{K} u_n(k)}{K} \geq w_n$, where $u_n(k)$ is defined as in Definition 2.15. By the definition of time-based debt, $\liminf_{K \to \infty} \frac{\sum_{k=1}^{K} u_n(k)}{K} \geq w_n$ is equivalent to $\liminf_{K \to \infty} \frac{r_n^{(1)}(K)}{K} \leq 0$. Thus, to establish the optimality of the largest time-based debt first policy, we only need to show that the set $A := \{z = [z_1, z_2, \ldots, z_N] | z_n \leq 0, \forall n\}$ is approachable under this policy.

Suppose that at the beginning of some interval, the average payoff is $x = [x_1, x_2, \ldots, x_N]$. We need only consider $x \notin A$. If $x \notin A$, at least one of $x_1, x_2, \ldots, x_N$ is strictly positive, and it is convenient to relabel the clients so that $x_1 \geq x_2 \geq \cdots \geq x_m > 0 \geq x_{m+1} \cdots \geq x_N$. The closest point in A to x is

$$y = [0, 0, \ldots, 0, x_{m+1}, x_{m+2}, \ldots, x_N].$$

The hyperplane passing through y and perpendicular to the line segment xy is $H := \{z | h(z) := \sum_{n=1}^{m} x_n z_n = 0\}$.

Let $\bar{x}$ be the payoff of this interval according to the largest time-based debt first policy. Also, let $\bar{w}_n$ be the number of times the AP transmits the packet from client n in the interval. We can express $\bar{x}$ as $\bar{x} = [w_1 - \bar{w}_1, w_2 - \bar{w}_2, \ldots, w_N - \bar{w}_N]$.

Since $h(x) = \sum_{n=1}^{m} x_n^2 > 0$, in order to show H separates x and $E[\bar{x}]$, it suffices to show that $h(\bar{x}) \leq 0$. We have:

$$h(\bar{x}) = \sum_{n=1}^{m} x_n(w_n - \bar{w}_n) \tag{2.15}$$

$$= \sum_{n=1}^{m-1} [(x_n - x_{n+1})(\sum_{k=1}^{n} w_k - \sum_{k=1}^{n} \bar{w}_k)] + x_m(\sum_{k=1}^{m} w_k - \sum_{k=1}^{m} \bar{w}_k). \tag{2.16}$$

Next we evaluate $\sum_{k=1}^{n} \bar{w}_k$ for each n. By the largest time-based debt first policy, the server will give priority according to the order $1, 2, \ldots, N$. Hence, $\sum_{k=1}^{n} \bar{w}_k$ is the number of transmissions the AP makes if only the subset $S_n = \{1, 2, \ldots, n\}$ of clients that is present in the system. In other words, we have

$$E[\sum_{k=1}^{n} \bar{w}_k] = T - I_{S_n}. \tag{2.17}$$

Now, according to the necessary condition stated in Lemma 2.13, we have $\sum_{k=1}^{n} w_k \leq T - I_{S_n} = \sum_{k=1}^{n} \bar{w}_k$, for all n. Further, $x_1 \geq x_2 \geq \cdots \geq x_m > 0$. Hence we have

$$E[h(\bar{x})] = \sum_{n=1}^{m-1}[(x_n - x_{n+1})(\sum_{k=1}^{n} w_k - E[\sum_{k=1}^{n} \bar{w}_k])]] + x_m(\sum_{k=1}^{m} w_k - E[\sum_{k=1}^{m} \bar{w}_k]) \qquad (2.18)$$

$$\leq 0. \qquad (2.19)$$

Thus, A is approachable under the largest time-based debt first policy by Theorem 2.19, which also implies that the largest time-based debt first policy is feasibility optimal. $\square$

Theorem 2.21   *The largest weighted-delivery debt first policy is also feasibility optimal.*

Proof. As in the previous proof, we again translate this policy into one for the single player game. Again, a round in the game corresponds to an interval consisting of T time slots in the model. The action a player, which is the AP, can take, is to decide the priority order of clients for that interval. However, in this case, we will define the payoff the player gets as the net change in the weighted-delivery debt. In other words, the payoff is an N-dimensional vector $v := [v_1, v_2, \ldots, v_N]$, where $v_n := (q_n - 1)/p_n$ if the AP delivers a packet for client n in the interval, or $v_n = q_n/p_n$ if not. The timely-throughput of a client n is at least q_n packets per interval if it approaches a non-positive weighted-delivery debt. Thus, we can prove that the largest weighted-delivery debt is optimal by showing that the set $A := \{z = [z_1, z_2, \ldots, z_N] | z_n \leq 0, \forall n\}$ is approachable.

Let $x = [x_1, x_2, \ldots, x_N]$ be the average payoff at the beginning of an interval. Again, we only need to evaluate the performance of the largest weighted-delivery debt first policy under the case $x \notin A$. For convenience, we relabel the clients so that $x_1 \geq x_2 \geq \cdots \geq x_m > 0 \geq x_{m+1} \geq \cdots \geq x_N$. The closest point in A to x is $y = [0, 0, \ldots, 0, x_{m+1}, x_{m+2}, \ldots, x_N]$. The hyperplane passing through y and perpendicular to the line segment xy is $H := \{z | h(z) := \sum_{n=1}^{m} x_n z_n = 0\}$.

Let π_n be the indicator function that the AP delivers a packet from client n, which is a random variable. The payoff of this interval is $\bar{x} = [(q_1 - \pi_1)/p_1, (q_2 - \pi_2)/p_2, \ldots, (q_N - \pi_N)/p_N]$.

By Theorem 2.19, the set A is approachable if H separates x and $E[\bar{x}]$. Since $h(x) = \sum_{n=1}^{m} x_n^2 > 0$, we only need to show $E[h(\bar{x})] \leq 0$ to complete the proof. We have:

$$h(\bar{x}) = \sum_{n=1}^{m} x_n \frac{q_n - \pi_n}{p_n} \qquad (2.20)$$

$$= \sum_{n=1}^{m-1}[(x_n - x_{n+1})(\sum_{k=1}^{n} \frac{q_k}{p_k} - \sum_{k=1}^{n} \frac{\pi_k}{p_k})] + x_m(\sum_{k=1}^{m} \frac{q_k}{p_k} - \sum_{k=1}^{m} \frac{\pi_k}{p_k}) \qquad (2.21)$$

$$= \sum_{n=1}^{m-1}[(x_n - x_{n+1})(\sum_{k=1}^{n} w_k - \sum_{k=1}^{n} \frac{\pi_k}{p_k})] + x_m(\sum_{k=1}^{m} w_k - \sum_{k=1}^{m} \frac{\pi_k}{p_k}) \text{ (since } w_k = \frac{q_k}{p_k}). \qquad (2.22)$$

Since $x_1 \geq x_2 \geq \cdots \geq x_m > 0$, it suffices to show $\sum_{k=1}^{n} w_k \leq E[\sum_{k=1}^{n} \frac{\pi_k}{p_k}]$, for every n. Recall that the necessary condition stated in Lemma 2.13 requires $\sum_{k=1}^{n} w_k \leq T - I_{S_n}$ for every n, to be feasible, where $S_n = \{1, 2, \ldots, n\}$. Thus, we only need to show $E[\sum_{k=1}^{n} \frac{\pi_k}{p_k}] = T - I_{S_n}$ to establish optimality, which is done in Lemma 2.22 below. $\square$

Lemma 2.22   *Under the priority order* $\{1, 2, \ldots, N\}$, $E[\sum_{k=1}^{n} \frac{\pi_k}{p_k}] = T - I_{S_n}$, *for* $n = 1, 2, \ldots, N$.

Proof. We prove this by induction. First consider the case $n = 1$. Since client 1 has the highest priority, its packet is delivered unless the AP fails in all the T attempts. Thus,

$$E[\frac{\pi_1}{p_1}] = \frac{Prob\{\text{the job of client 1 is accomplished}\}}{p_1} = \frac{1 - (1 - p_1)^T}{p_1}. \tag{2.23}$$

On the other hand, we also have:

$$I_{S_1} = \sum_{t=1}^{T-1} Prob\{\text{number of idle time slots is at least } t\} \tag{2.24}$$

$$= \sum_{t=1}^{T-1} Prob\{\text{the packet for client 1 is delivered in at most } T - t \text{ transmissions}\} \tag{2.25}$$

$$= \sum_{t=1}^{T-1} (1 - (1 - p_1)^{T-t}) = T - \frac{1 - (1 - p_1)^T}{p_1}. \tag{2.26}$$

This gives us $E[\frac{\pi_1}{p_1}] = T - I_{S_1}$, and the lemma holds for the case $n = 1$.

Assume that $E[\sum_{k=1}^{n} \frac{\pi_k}{p_k}] = T - I_{S_n}$ holds for all $n \leq m$. Consider the case $n = m + 1$. Since the client $m + 1$ has the lowest priority among clients $\{1, 2, \ldots, m + 1\}$, its packet is transmitted only after all packets from client 1 through client m are delivered. Let L_{S_m} be the number of time slots left after the AP delivers packets from the first m clients. Note that L_{S_m} is a random variable. Suppose $L_{S_m} = \sigma$. From client m's point of view, this is as if the length of the interval is σ, and it is the client with the highest priority. Using the above analysis, we have:

$$E[\pi_{m+1}|L_{S_m} = \sigma] = 1 - (1 - p_{m+1})^{\sigma}. \tag{2.27}$$

Moreover, given that $L_{S_m} = \sigma$, the number of time slots left in the interval after the packet of client $m + 1$ is delievered can be expressed as

$$E[L_{S_{m+1}}|L_{S_m} = \sigma] = \sigma - \frac{1 - (1 - p_{m+1})^{\sigma}}{p_{m+1}} \tag{2.28}$$

$$= \sigma - E[\frac{\pi_{m+1}}{p_{m+1}}|L_{S_m} = \sigma], \tag{2.29}$$

for all σ. Finally, as $I_{S_m} := E[L_{S_m}]$, we have:

$$E[\sum_{k=1}^{m+1} \frac{\pi_k}{p_k}] = E[\sum_{k=1}^{m} \frac{\pi_k}{p_k}] + E[E[\frac{\pi_{m+1}}{p_{m+1}} | L_{S_m}]] \tag{2.30}$$

$$= T - I_{S_m} + E[L_{S_m} - E[L_{S_{m+1}} | L_{S_m}]] \tag{2.31}$$

$$= T - I_{S_m} + I_{S_m} - I_{S_{m+1}} = T - I_{S_{m+1}}. \tag{2.32}$$

By induction, the lemma therefore holds for all n.   □

A final remark is that since both policies fulfill every set of clients that satisfies the necessary condition in Lemma 2.13, this condition is also sufficient for feasibility.

Theorem 2.23   *A set of clients is feasible if and only if*

$$\sum_{n \in S} w_n \leq T - I_S, \tag{2.33}$$

holds for every subset S.

2.5   SIMULATION RESULTS

We now present some simulation results on VoIP traffic to demonstrate Theorem 2.23 and the performance of the two largest debt first policies. We follow the G.711 codec, which is a ITU-T standard for audio compression, in deciding parameters for traffic with QoS constraints. G.711 generates data at 64 kbps. With a 20 *ms* packetization interval, this results in a 160 Bytes VoIP packet. We use IEEE 802.11b as the underlying MAC protocol, whose transmission rate can be as high as 11 Mb/s. Some details of parameters are given in the table below. Under this setting, the total transmission time for a CF-POLL packet and a Data packet is slightly smaller than 610 μs, allowing 32 time slots in a 20 *ms* period. All the results in this section are the average over 100 runs.

We have implemented the two largest debt first policies, the largest time-based debt first policy and the largest weighted-delivery debt first policy, on ns-2, and compared them against the naive approach using IEEE 802.11 DCF. To evaluate the performance of different mechanisms, we define a *deadline miss ratio (DMR)* function. Let d_n be the timely-throughput achieved by the client n at some given time, defined as the number of packets delivered divided by the number of intervals up till that time. The *deadline miss ratio of client n* is defined as

$$DMR_n = \begin{cases} q_n - d_n, & \text{if } q_n > d_n, \\ 0, & \text{otherwise,} \end{cases}$$

and the *deadline miss ratio of the system* is defined as the sum of the deadline miss ratios of all clients.

One may argue that the comparison between largest debt policies and IEEE 802.11 DCF is not fair. By using IEEE 802.11 PCF, our policies avoid additional overheads, including the time spent on backing off and the risk of packet collision, which are inevitable in DCF. To make the comparison fair, we therefore also implement a *random policy*, which is also based on PCF, that assigns priorities randomly at the beginning of each period.

Table 2.1: Simulation Setup

Packetization interval	20 *ms*
Payload size per packet	160 Bytes
Transmission data rate	11 Mb/s
SIFS	10 μs
PIFS	30 μs
DIFS	40 μs

We consider two groups of clients, group A and group B. Clients in group A carry more important messages and require a 0.99 timely-throughput, while clients in group B require a 0.8 timely-throughput. The channel reliability of the nth client in both groups is assumed to be $(60 + n)\%$. Using Theorem 2.23, it can be shown that a set of 11 group A clients and 12 group B clients is feasible, but a set of 12 group A clients and 12 group B clients is not.

We first run simulations of the four different policies, namely, the two largest debt first policies, the random policy based on PCF, and the DCF mechanism, on a set of 12 group A clients and 11 group B clients. Figure 2.6 presents the deadline miss ratios of different policies. The deadline miss ratios of the two largest debt first policies converge to zero over time, showing that both policies fulfill the set of clients. The largest weighted-delivery debt first policy has a better performance over the largest time-based debt first policy since it converges faster. This is because the largest weighted-delivery debt first policy uses the feedback information from MAC to count the actual number of packets delivered for each client. This provides a better estimate of whether a client requires more transmission opportunities. The largest time-based debt first policy, on the other hand, uses the more indirect approach by counting the number of times a client transmits without regard to the number of successful deliveries, which gives a slower convergence rate. The largest weighted-delivery debt first policy is definitely the preferred choice.

Both the other two policies do not satisfy all clients. The deadline miss ratio of the random policy remains approximately 1. The random policy cannot be feasibility optimal since it gives each client equal priorities in the long term, regardless of the required delivery ratio of each client. However, clients with more important data should be granted more transmission opportunities than others. Failing to take this factor into account makes the random policy not feasibility optimal. Meanwhile, the DCF mechanism has a much higher deadline miss ratio. This is due to the lack of awareness of delay constraints in DCF. When a client puts a packet in the transmission queue, the packet cannot be removed until it is transmitted or dropped by the MAC. Thus, when the packet generation rate exceeds the packet outgoing rate, the queuing delay gets even larger, resulting in a large deadline miss ratio.

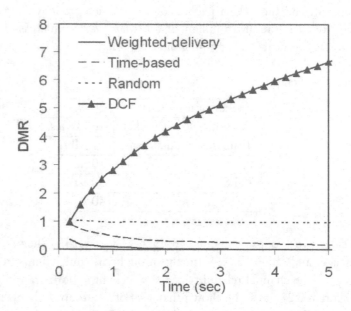

Figure 2.6: Deadline miss ratio of a feasible set

To illustrate the accuracy of Theorem 2.23, we run simulations on a set of 12 group A clients and 12 group B clients, which is known to be infeasible by Theorem 2.23. Figure 2.7 shows the results. It can be seen that all the four policies result in non-zero deadline miss ratios, confirming the infeasibility of this set. It can also be seen that, as in the case of the feasible set, the two largest debt first policies result in the least deadline miss ratios. This result suggests that our proposed policies still work well even when link qualities are not high enough to satisfy the clients. Since the link quality does vary with time in wireless networks, it is likely that the network may suffer from temporary link quality downgrades from time to time. The ability to provide good service under these downgrades is essential to the robustness of a system. In this respect, our policies appear to be robust.

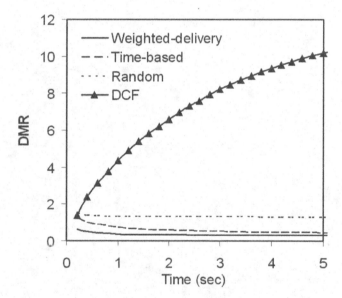

Figure 2.7: Deadline miss ratio of an infeasible set

CHAPTER 3

Admission Control

In Chapter 2, we derived the necessary and sufficient condition for feasibility of the quality of service in a base case. Admission control therefore consists simply of evaluating this necessary and sufficient condition. However, as stated, the condition requires testing the inequalities $\sum_{n \in S} w_n \leq T - I_S$ for every subset S, which results in exponentially many tests (2^N) and may not be computationally efficient. Also, the base case assumes that the channel reliability p_n for each client does not change over time. In other words, the base case neglects fading effects. In this chapter, we begin by establishing a polynomial time algorithm for admission control for the base case. We then consider the case of systems with fading channels.

3.1 AN EFFICIENT ALGORITHM WHEN PACKET GENERATION IS PERIODIC

In Chapter 2, we have developed a condition for feasibility, as stated in Theorem 2.23. It however entails evaluating inequalities for every subset of N clients, and thus results in 2^N tests, which is exponentially many tests in N. In this section, we begin by showing that we only need to evaluate a total of N conditions to determine feasibility. We thus obtain a polynomial-time algorithm for admission control.

Theorem 3.1 *Order the clients so that $q_1 \geq q_2 \geq \cdots \geq q_N$. Let S_k be the subset $\{1, 2, \ldots, k\}$. The system is then feasible if and only if $\sum_{n \in S_k} w_n + I_{S_k} \leq T$, for all $1 \leq k \leq N$.*

Proof. It is obvious that the above condition is necessary for feasibility. We only need to show that it is sufficient.

Consider any infeasible system. Define a *minimal infeasible set* of this system as a subset S for which $\sum_{n \in S} w_n + I_S > T$, but for each $S' \subsetneq S$, $\sum_{n \in S'} w_n + I_{S'} \leq T$. For every infeasible system, there must exist at least one minimal infeasible set S. Fix S, and let m be the largest element in S; that is, $m := \min\{k | S \subseteq S_k\}$. We prove that $\sum_{n \in S_m} w_n + I_{S_m} > T$.

If $S = S_m$, then we are done. Otherwise, let l be the largest element in $S_m \setminus S$. We want to show that $\sum_{n \in S \cup \{l\}} w_n + I_{S \cup \{l\}} \geq \sum_{n \in S} w_n + I_S$.

Consider a scheduling policy η that only transmits packets for clients in $S \setminus \{m\}$ and fulfills them. Such a policy exists because S is a minimal infeasible set, whence the subset $S \setminus \{m\}$ is feasible. We expand η by making it schedule transmissions for client m in an interval only after all packets for clients in $S \setminus \{m\}$ are delivered, and schedule transmissions for client l only after the packet for client

m is delivered. Under this policy, the expected amount of time that the AP spends transmitting the packet for client l in an interval is $I_S - I_{S \cup \{l\}}$, and so the timely-throughput of client l is $p_l(I_S - I_{S \cup \{l\}})$ by Lemma 2.5. Also, the timely-throughput of client m is strictly less than q_m, since the set S is infeasible, and all clients in $S \setminus \{m\}$ are fulfilled under this policy. Further, the timely-throughput of client l is no larger than that of client m, as it is scheduled only after the packet for client m is delivered. Thus, $p_l(I_S - I_{S \cup \{l\}}) < q_m$. We now have

$$\left(\sum_{n \in S \cup \{l\}} w_n + I_{S \cup \{l\}} \right) - \left(\sum_{n \in S} w_n + I_S \right) = \frac{q_l}{p_l} - (I_S - I_{S \cup \{l\}}) \tag{3.1}$$

$$= [q_l - p_l(I_S - I_{S \cup \{l\}})]/p_l \tag{3.2}$$

$$> (q_l - q_m)/p_l \geq 0, \tag{3.3}$$

where the last inequality holds because $q_l \geq q_m$. Hence, $\sum_{n \in S \cup \{l\}} w_n + I_{S \cup \{l\}} > \sum_{n \in S} w_n + I_S > T$.

If $S \cup \{l\} = S_m$, then we are done. Otherwise, we select l' to be the largest element in $S_m \setminus (S \cup \{l\})$. We expand η by making it schedule transmissions for client m in an interval only after all packets for clients in $S \setminus \{m\}$ are delivered, and schedule transmissions for client l (or client l') only after the packet for client m (or the packet for client l, respectively) is delivered. Under this policy, the expected number of time slots that the AP spends on client l' is $I_{S \cup \{l\}} - I_{S \cup \{l,l'\}}$, and thus the timely-throughput of client l' is $p_{l'}(I_{S \cup \{l\}} - I_{S \cup \{l,l'\}})$. Further, the timely-throughput of client l' is no larger than that of client l, which in turn is strictly less than q_m. We now have

$$\left(\sum_{n \in S \cup \{l,l'\}} w_n + I_{S \cup \{l,l'\}} \right) - \left(\sum_{n \in S \cup \{l\}} w_n + I_{S \cup \{l\}} \right) = \frac{q_{l'}}{p_{l'}} - (I_{S \cup \{l\}} - I_{S \cup \{l,l'\}}) \tag{3.4}$$

$$= [q_{l'} - p_{l'}(I_{S \cup \{l\}} - I_{S \cup \{l,l'\}})]/p_{l'} \tag{3.5}$$

$$> (q_{l'} - q_m)/p_{l'} \geq 0. \tag{3.6}$$

Hence, $\sum_{n \in S \cup \{l,l'\}} w_n + I_{S \cup \{l,l'\}} > \sum_{n \in S} w_n + I_S > T$.

If $S \cup \{l, l'\} = S_m$, then we are done. Otherwise, we select l'' to be the largest element in $S_m \setminus (S \cup \{l, l'\})$ and repeat the above procedure. By induction, we establish that $\sum_{n \in S_m} w_n + I_{S_m} > T$. $\qquad \square$

In addition to reducing the number of needed tests to N, this theorem also helps improve the efficiency of evaluating each test. In each test, we need to obtain the values of $\sum_{n \in S_m} w_n$ and I_{S_m}, both of which require more than a constant computation time. However, by using the fact that $S_m = S_{m-1} \cup \{m\}$, we can recursively compute these values very efficiently and improve complexity.

Obtaining $\sum_{n \in S_m} w_n$ is easy since it equals $\sum_{n \in S_{m-1}} w_n + w_m$, and requires only one addition operation given the value of $\sum_{n \in S_{m-1}} w_n$. The computation of I_{S_m} is more complicated. Let $g_{S_m}(t)$ be the probability that all packets in S_m are delivered at t, resulting in $T - t$ idle time slots. I_{S_m}, being the expected number of idle time slots in an interval, is $\sum_{i=1}^{T}(T - i)g_{S_m}(t)$. Consider the value of $g_{S_m}(t)$:

$$g_{S_m}(t) = Prob(\text{all packets in } S_m \text{ are delivered at } t) \tag{3.7}$$

$$= \sum_{i=1}^{T} Prob(\text{all packets in } S_{m-1} \text{ are delivered at time } t - i, \tag{3.8}$$

and clients m takes i time slots to succeed)

$$= \sum_{i=1}^{T} g_{S_{m-1}}(t - i)[p_m(1 - p_m)^{i-1}]. \tag{3.9}$$

Thus, the vector of $[g_{S_m}(i)]$ is indeed the convolution of the vectors of $[g_{S_{m-1}}(i)]$ and $[p_m(1 - p_m)^{i-1}]$, which can be computed in $O(T^2)$ time by brute force, or $O(T \log T)$ by using the Fast Fourier Transform algorithm.

A complete algorithm for deciding whether a set of clients is feasible is given in Algorithm 1. The complexity of the algorithm is $O(NT^2)$ or $O(NT \log T)$, depending on the implementation of convolution.

Algorithm 1 IsFeasible

1: Assume clients are sorted so that $q_1 \geq q_2 \geq \cdots \geq q_N$
2: $totalW \leftarrow 0$
3: $[g_{S_0}(i)] \leftarrow [1, 0, \cdots, 0]$
4: **for** $m = 1$ to N **do**
5:     $totalW \leftarrow totalW + \frac{q_m}{p_m}$
6:     $[g_{S_m}(i)] \leftarrow [g_{S_{m-1}}(i)] * [p_m(1 - p_m)^{i-1}]$
7:     $totalI \leftarrow \sum_{i=1}^{T}(T - i)g_{S_m}(i)$
8:     **if** $totalW + totalI > T$ **then**
9:         **return** Infeasible
10:     **end if**
11: **end for**
12: **return** Feasible

3.2   ADMISSION CONTROL UNDER FADING CHANNELS

We now address how to extend the feasibility conditions to fading channels. We first introduce a model for fading channels. We model the channel condition as a stationary, irreducible Markov process with a finite set of channel states $\mathcal{C}$. The average probability that channel state c occurs is f_c, with the channel state remaining constant within each interval. Under channel state c, the link reliability between the AP and client n is $p_{c,n}$, by which we mean that a packet transmitted by the AP for client n is delivered with probability $p_{c,n}$.

One intuitive approach is to decouple the channel states. The AP assigns a timely-throughput requirement $q_{c,n}$ for each channel state c and client n, with $\sum_{c \in C} f_c q_{c,n} \geq q_n$. Also, for each channel state c, the assigned throughput requirements must be feasible under that channel state, that is, $\sum_{n \in S} \frac{q_{c,n}}{p_{c,n}} \leq T - E[I_{c,S}]$ for all $S \subseteq \{1, 2, \ldots, N\}$, where $I_{c,S}$ is the expected number of time slots that the AP is forced to stay idle in an interval under channel state c for any work conserving scheduling policy, given that the AP only transmits packets for the subset S of clients. More formally, we therefore seek a matrix $Q = [q_{c,n}]$ that solves the following linear programming problem:

$$\text{Max} \sum_{n=1}^{N} \sum_{c \in C} f_c q_{c,n} \tag{3.10}$$

$$\text{s.t.} \sum_{c \in C} f_c q_{c,n} \geq q_n, \forall n \tag{3.11}$$

$$\sum_{n \in S} \frac{q_{c,n}}{p_{c,n}} \leq T - E[I_{c,S}], \forall c, \forall S \subseteq \{1, 2, \cdots, N\}. \tag{3.12}$$

Theorem 3.2 *A set of clients is feasible if and only if there exists a matrix Q that solves the above linear programming problem.*

After determining the matrix Q, we can apply the two largest debt first policies in Chapter 2 for each channel state individually to fulfill a feasible system. However, solving the above linear programming problem involves in exponentially many constraints, and is computationally complicated. Further, the efficient algorithm in Chapter 3.1 cannot be applied to reduce the number of constraints. The reason is that Algorithm 1 sorts all clients according to their values of q_n in the first step, which requires knowing the values of q_n. In contrast, we do not know the ordering of $q_{c,n}$ in advance when solving the linear programming problem, which makes Algorithm 1 inapplicable. In the next chapter, we will demonstrate that there exists a simple online scheduling policy that fulfills every feasible system with fading channels. Thus, even though characterization of feasibility of a set of fading clients has high complexity, it turns out that there is a simple policy that is feasibility-optimal.

CHAPTER 4

Scheduling Policies

We have proposed two largest debt first scheduling policies in Chapter 2. The two policies are feasibility optimal in the base case. In this chapter, we discuss the problem of packet scheduling for more general cases than in the base case. We will consider the case of stochastic packet arrivals, client-dependent deadlines, time varying channels, and rate adaptation whereby clients can reduce their transmit rate to improve reliability.

4.1 AN EXTENDED SYSTEM MODEL

We first introduce an extended system model for more general cases.

Consider a wireless system with N clients, $\{1, 2, \ldots, N\}$, and one access point (AP). Similar to the basic system model in Section 2.1, time is slotted with slots denoted by $t \in \{0, 1, 2, \ldots\}$. Time slots are further grouped into *intervals* $[kT, (k+1)T)$, with interval length T. The AP is in charge of scheduling all transmissions, and we can consider both downlink transmissions and uplink transmissions in the same way as in Section 2.1.

However, suppose now that packets for each client are generated probabilistically at the beginning of each interval, at time slots $\{0, T, 2T, \ldots\}$ with no more than one packet per client. We model the packet generations as a stationary, irreducible Markov process with finite state. The average probability that packets are generated for the subset S of clients is $R(S)$. Packet generations can be dependent between clients, and packet generations in an interval can depend on other intervals.

Clients may have different relative deadlines. Client n hass a delay requirement τ_n measured in slots, with $\tau_n \leq T$. If the packet for client n is not delivered by the τ_n^{th} time slot of the interval, the packet expires and is discarded.

Next we turn to the model of the channel. We consider an unreliable, heterogeneous, and time-varying channel model. We model the channel condition as a stationary, irreducible Markov process with a finite set of channel states $\mathcal{C}$. The average probability that channel state c occurs is f_c, with the channel state remaining constant within each interval.

We consider systems both with rate adaptation and without. When rate adaptation is not available, that is, when all packets are transmitted at a fixed rate, the AP can make exactly one transmission in each time slot. Under channel state c, the link reliability between the AP and client n is $p_{c,n}$, by which we mean that a packet transmitted by the AP for client n is delivered with probability $p_{c,n}$. On the other hand, when the system uses rate adaptation, the channel states describe the maximal rates that can be supported between the AP and clients, which in turn decide the service times for transmissions. Under channel state c, it takes $s_{c,n}$ time slots to make an error free transmission to client n.

The channel state and the packet arrivals in an interval are assumed to be independent of each other. We also assume that the AP has knowledge of channel state.

The performance of a client is measured in terms of its *timely-throughput*, defined as in Section 2.1. Each client n requires a timely-throughput of at least $q_n > 0$ packets per interval. Since, on average, there are $\sum_{S:n\in S} R(S)$ packets for client n per interval, this timely-throughput bound can also be interpreted as a delivery ratio requirement of $\frac{q_n}{\sum_{S:n\in S} R(S)}$.

We slightly modify the technical definition of timely-throughput.

Definition 4.1   A set of clients, $\{1, 2, \ldots, N\}$ is *fulfilled* under a scheduling policy η, if for every $\epsilon > 0$,

$$Prob\{\frac{d_n(t)}{t/T} > q_n - \epsilon, \text{ for every } n\} \to 1, \text{ as } t \to \infty,$$

where $d_n(t)$ is the number of packets delivered to client n up to time t.[1]

Since the overall system can be viewed as a controlled Markov chain, we have:

Lemma 4.2   *For any set of clients that can be fulfilled, there exists a stationary randomized policy that fulfills the clients, which uses a probability distribution based only on the channel state, the set of undelivered packets, and the number of time slots remaining in the system (and not depending on any events in past intervals), according to which it randomly chooses an undelivered packet to transmit, or stays idle.*

Definition 4.3   A set of clients is *feasible* if there exists some scheduling policy that fulfills it.

Definition 4.4   We call the region in N-space formed by vectors $[q_n]$ for which the clients are feasible as the *feasible region*.

Lemma 4.5   *The feasible region is a convex set.*

Proof.   Let $[q_n]$ and $[q_n']$ be two vectors in the feasible region. Then, by Lemma 4.2, there exist stationary randomized policies η and η' that fulfill the two vectors, respectively. Then, the policy that randomly picks one of the two policies, with η being chosen with probability α, at the beginning of each interval, fulfills the vector $[\alpha q_n + (1 - \alpha)q_n']$. Further, since q_n and q_n' are both larger than 0 for each n, $\alpha q_n + (1 - \alpha)q_n' > 0$ for all n. Thus, the vector $[\alpha q_n + (1 - \alpha)q_n']$ also falls in the feasible region. □

---

[1]In other words, we require that $(\frac{d_n(t)}{t/T} - q_n)^+$ converges to zero in probability, for every n. In contrast, previous chapters required that $(\frac{d_n(t)}{t/T} - q_n)^+$ converges to zero almost surely.

Note that if $[q_n]$ is feasible, then so is $[q'_n]$, where $0 < q'_n \le q_n$, for all n.

Definition 4.6 $[q_n]$ is *strictly feasible* if there exists some $\alpha \in (0, 1)$ such that $[q_n/\alpha]$ is still feasible.

Finally, we modify the concept of feasibility optimal policies.

Definition 4.7   A scheduling policy η is *feasibility optimal* if it fulfills every set of clients that is strictly feasible.

Note that in this chapter, we only require a feasibility optimal policy to fulfill every *strictly feasible* system. In other words, a feasibility optimal policy only needs to fulfill vectors $[q_n]$ that are within the interior of the feasible region. On the other hand, in Chapter 2, we required a feasibility optimal to fulfill every *feasible* system, which includes vectors $[q_n]$ that are on the boundary of the feasible region.

4.2   A FRAMEWORK FOR DETERMINING SCHEDULING POLICIES

We now describe a more general class of policies that is feasibility optimal. We start by extending the concept of "debt."

Definition 4.8   A variable $r_n(k)$, whose value is determined by the past history of the client n up to the kth interval, or time slot kT, is called a *pseudo-debt* if:

1. $r_n(0) = 0$, for all n.

2. $r_n(k + 1) = r_n(k) + z_n(q_n) - \mu_n(k)$, where $\mu_n(k)$ is a non-negative and bounded random variable whose value is determined by the behavior of client n, and $z_n(q_n)$ is a strictly positive number, with $z_n(\cdot)$ a strictly increasing linear function. Further, $\mu_n(k) = 0$ if the AP does not transmit any packet for client n.

3. The set of clients is fulfilled if and only if $Prob\{\frac{r_n(k)}{k} < \varepsilon\} \to 1$, as $k \to \infty$, for all n and all $\varepsilon > 0$.

In the following example, we illustrate that both the time-based debt and the weighted-delivery debt are pseudo-debts under a static channel model.

Example 4.9   At the beginning of each interval, the time-based debt $r_n^{(1)}(k)$ increases by $w_n = \frac{q_n}{p_n}$, and decreases by the number of time slots that the AP has transmitted packets for client n during the interval. Lemma 2.5 shows that condition (3) is satisfied.

Similarly, $r_n^{(2)}(k)$, the weighted-delivery debt is also a special case, where $z_n(q_n) = \frac{q_n}{p_n}$, and $\mu_n(k)$ is the ratio (number of packets delivered for n in interval k)/p_n. It satisfies condition (3) by definition.

We can also consider the *delivery debt* $r_n^{(3)}(k) := q_n k - d_n(kT)$, where $d_n(t)$ is the number of delivered packets for client n up to time slot t. For the delivery debt, $z_n(q_n) := q_n$, and $\mu_n(k)$ is the number of packets delivered for client n in interval k. Thus, the delivery debt is also a pseudo-debt. ☐

We can also define the *feasible region for debt* as the set of $[z_n(q_n)]$ such that the corresponding $[q_n]$ is feasible. Since $z_n(q_n)$ is a linear function of q_n and the feasible region for $[q_n]$ is a convex set (Lemma 4.5), the feasible region for $[z_n(q_n)]$ is also a convex set.

Using the concept of pseudo-debt, we can establish a sufficient condition for feasibility optimality. The proof is based on:

Theorem 4.10   Lyapunov Drift Theorem   *Let $L(t)$ be a non-negative Lyapunov function. Suppose there exists some constant $B > 0$ and a non-negative $f(t)$ adapted to the past history of the system such that:*

$$E\{L(t+1) - L(t)|H_t\} \leq B - \epsilon f(t),$$

for all t, where H_t is the system history up to time t. Then: $\limsup_{t\to\infty} \frac{1}{t} \sum_{i=0}^{t} E\{f(i)\} \leq B/\epsilon.$ ☐

The Lyapunov Drift Theorem is an important tools in designing scheduling policies and proving their stability for queueing systems. Interested readers can find more in-depth discussions on this theorem and its applications to queueing systems in Appendix A. The following theorem is similar to the design of *Maximum Weight* policy for queueing systems, which is also discussed in Appendix A.

Theorem 4.11   *Let $r_n(k)$ be a pseudo-debt. A policy that maximizes the* payoff function

$$\sum_{n=1}^{N} E\{r_n(k)^+ \mu_n(k)|c_k, S_k, [r_m(k)]\} \tag{4.1}$$

in each interval is feasibility optimal, where c_k denotes the channel state in the kth interval, S_k is the set of packets that arrive at the AP at the beginning of the kth interval, and $x^+ := \max\{x, 0\}$.

Proof. Define $L(k) = \frac{1}{2} \sum_{n=1}^{N} r_n(k)^2$. Since $r_n(k+1) = r_n(k) + z_n(q_n) - \mu_n(k)$,

$$\Delta(L(k)) := E\{L(k+1) - L(k)|c_k, S_k, [r_m(k)]\} \tag{4.2}$$

$$= E\{\frac{1}{2} \sum_{n=1}^{N} r_n(k+1)^2 - \frac{1}{2} \sum_{n=1}^{N} r_n(k)^2 |[r_m(k)]\} \tag{4.3}$$

$$= E\{\sum_{n=1}^{N} r_n(k)[z_n(q_n) - \mu_n(k)]|c_k, S_k, [r_m(k)]\} + B(k), \tag{4.4}$$

where $B(k) := E\{\frac{1}{2} \sum_{n=1}^{N} [z_n(q_n) - \mu_n(k)]^2 | c_k, S_k, [r_m(k)]\}$. Then $B(k) \leq B$, for all k, for some B. Hence for any policy:

$$\Delta(L(k)) \leq E\{\sum_{n=1}^{N} r_n(k)[z_n(q_n) - \mu_n(k)]|c_k, S_k, [r_m(k)]\} + B. \tag{4.5}$$

Suppose $[q_n]$ is strictly feasible. The vector $[z_n(q_n)]$ is thus an interior point of the feasible region (for debt), and there therefore exists some $\alpha \in (0, 1)$ such that $[z_n(q_n)/\alpha]$ is also in the feasible region. Let $z_{min} := \min\{z_1(q_1), z_2(q_2), \ldots, z_N(q_N)\}$. The N-dimensional vector $[z_{min}]$ whose elements are all z_{min} falls in the feasible region. Since the feasible region is convex, the vector $\alpha[z_n(q_n)/\alpha] + (1-\alpha)[z_{min}] = [z_n(q_n) + (1-\alpha)z_{min}]$ is also in the feasible region.

By Lemma 4.2, there exists a stationary randomized policy η' that fulfills the set of clients with timely-throughput bounds for the vector $[z_n(q_n) + (1-\alpha)z_{min}]$. Since both channel states and packet arrivals can be modeled by a finite-state Markov chain, there exists some constant M such that the expected timely-throughput of client n under η' is at least q'_n in any M consecutive intervals regardless of the system history prior to the M consecutive intervals, where q'_n is chosen so that $z_n(q'_n) \geq z_n(q_n) + (1-\alpha)z_{min} - \frac{(1-\alpha)z_{min}}{2}$. In other words, let $\mu'_n(k)$ be the decrease in the pseudo-debt for client n under η' during interval k, then there exists M such that

$$E\{\frac{\sum_{i=k}^{k+M-1} \mu'_n(i)}{M}|c_k, S_k, [r_m(k)]\} \tag{4.6}$$

$$\geq z_n(q_n) + (1-\alpha)z_{min} - \frac{(1-\alpha)z_{min}}{2} \tag{4.7}$$

$$= z_n(q_n) + \frac{(1-\alpha)z_{min}}{2}, \tag{4.8}$$

for all k.

Let η be a policy that maximizes the payoff function (4.1), for all k. Then defining $\mu_n(k)$ and $r_n(k)$ as the decreases resulting from policy η and the pseudo-debt policy, we have:

$$\sum_{n=1}^{N} E\{r_n(k)^+ \mu_n(k)|c_k, S_k, [r_m(k)]\} \tag{4.9}$$

$$\geq \sum_{n=1}^{N} E\{r_n(k)^+ \mu_n'(k)|c_k, S_k, [r_m(k)]\}. \tag{4.10}$$

We can assume without loss of generality that the policy does not work on any client n with $r_n(k) \leq 0$, that is, $\mu_n(k) = 0$ if $r_n(k) \leq 0$.[2] From (4.5), we obtain that, under η:

$$E\{\frac{L(k+M)}{M} - \frac{L(k)}{M}|c_k, S_k, [r_m(k)]\} \tag{4.11}$$

$$= E\{\frac{\sum_{i=k}^{k+M-1} \Delta(L(i))}{M}|c_k, S_k, [r_m(k)]\} \tag{4.12}$$

$$\leq E\{\sum_{i=k}^{k+M-1} \sum_{n=1}^{N} r_n(i)^+[z_n(q_n) - \mu_n(i)]|c_k, S_k, [r_m(k)]\}/M + B \tag{4.13}$$

$$\leq E\{\sum_{i=k}^{k+M-1} \sum_{n=1}^{N} r_n(i)^+[z_n(q_n) - \mu_n'(i)]|c_k, S_k, [r_m(k)]\}/M + B \tag{4.14}$$

$$\leq E\{\sum_{n=1}^{N} r_n(k)^+[z_n(q_n) - \sum_{i=k}^{k+M-1} \frac{\mu_n'(i)}{M}]|c_k, S_k, [r_m(k)]\} + B + A \tag{4.15}$$

$$\leq -\sum_{n=1}^{N} r_n(k)^+ \frac{(1-\alpha)z_{min}}{2} + B + A, \tag{4.16}$$

where A is a constant, as $z_n(q_n)$, $\mu_n'(i)$, and $r_n(i) - r_n(k)$ are all bounded for all n and $i \in [k, k + M - 1]$.

Let $\epsilon := (1 - \alpha)z_{min}/2$. Let $\hat{L}(k) := L(kM)/M$. By the above discussion, we have $E\{\hat{L}(k + 1) - \hat{L}(k)|\text{system history up to the } (kM - 1)^{th} \text{ interval}\} \leq A + B - \epsilon \sum_{n=1}^{N} r_n(kM)^+$. By Theorem 4.10,

$$\limsup_{k \to \infty} \frac{1}{k} \sum_{i=0}^{k} E\{\sum_{n=1}^{N} r_n(kM)^+\} \leq (A + B)/\epsilon. \tag{4.17}$$

Finally, since $|\sum_{n=1}^{N} r_n(kM + M)^+ - \sum_{n=1}^{N} r_n(kM)^+|$ is bounded for all k, (4.17) implies that $\frac{1}{k} E\{\sum_{n=1}^{N} r_n(kM)^+\} \to 0$ as $k \to \infty$, as shown in Lemma 4.12 below. This shows that $\frac{r_n(kM)^+}{k}$ converges to 0 in probability for all n. As $|r_n(kM)^+ - r_n(i)^+|$ is bounded for all $i \in [kM, kM + M]$, and M is also bounded, we have $\frac{r_n(k)^+}{k}$ converges to 0 in probability for all n. Hence, η is feasibility optimal. □

---

[2]Since a policy cannot lose its feasibility optimality by doing more work, this assumption is not restrictive.

Lemma 4.12   *Let $f(t)$ be a non-negative function such that $|f(t+1) - f(t)| \leq M$, for some $M > 0$, for all t. If $\limsup_{t \to \infty} \frac{1}{t} \sum_{i=0}^{t} f(i) \leq B$, for some constant B, then $\lim_{t \to \infty} \frac{1}{t} f(t) = 0$.*

Proof. We prove by contradiction. Suppose $\limsup_{t \to \infty} \frac{1}{t} f(t) > \delta$, for some $\delta > 0$. Thus, $f(t) > t\delta$ infinitely often. Suppose $f(t) > t\delta$ for some t. Since $|f(t) - f(t-1)| < M$, we have $f(t-1) > t\delta - M$. Similarly, $f(t-2) > t\delta - 2M$, $f(t-3) > t\delta - 3M, \ldots, f(t - \lfloor t\delta/M \rfloor) > t\delta - \lfloor t\delta/M \rfloor M \geq 0$. Summing over these terms gives: $\sum_{i=t-\lfloor t\delta/M \rfloor}^{t} f(i) > \frac{t\delta \lfloor t\delta/M \rfloor}{2}$, and thus, $\sum_{i=0}^{t} \frac{1}{t} f(i) > \frac{\delta \lfloor t\delta/M \rfloor}{2}$. Since $f(t) > t\delta$ infinitely often, $\limsup_{t \to \infty} \sum_{i=0}^{t} \frac{1}{t} f(i) = \infty$, which is a contradiction. $\qquad \square$

Theorem 4.11 suggests a more general procedure for designing feasibility optimal scheduling policies. To design a scheduling policy in a particular scenario, we need to choose an appropriate pseudo-debt and obtain a policy to maximize the payoff function. Maximizing the payoff function is, however, in general, difficult. Nevertheless, in some special cases, evaluating the payoff function gives us simple feasibility optimal policies, or, at least, some insights into designing a reasonable heuristic, as long as we choose the correct pseudo-debt. In the following sections, we demonstrate the utility of this approach.

4.3   SCHEDULING OVER UNRELIABLE FADING CHANNELS

We now consider the case when all clients have the same delay bound, $\tau_n \equiv T$. We also suppose that the AP uses a fixed transmission rate and has instant knowledge of link reliabilities, which are time-varying. We propose an on-line scheduling policy and prove that it is feasibility optimal.

We use the delivery debt, $r_n^{(3)}(k)$, of Example 4.9. Thus, $\mu_n(k)$ is the number of packet deliveries for client n in the kth interval.

We call our proposed policy the *joint debt-channel policy*. Suppose that at the beginning of an interval, the delivery debt vector is $[r_n^{(3)}(k)]$, the channel state is c, and the set of arrived packets is S. The joint debt-channel policy prioritizes all clients who have packet arrivals and positive $r_n^{(3)}(k)$ according to $r_n^{(3)}(k)p_{c,n}$, where clients with larger $r_n^{(3)}(k)p_{c,n}$ get higher priorities. Algorithm 2 formally describes this policy. The computational complexity of this policy is $O(N \log N)$ per interval.

Algorithm 2 Joint Debt-Channel Policy

1: **for** $n = 1$ to N **do**
2: $\quad r_n^{(3)}(k) \leftarrow q_n k - d_n(kT)$, for all n
3: **end for**
4: Sort clients with packet arrivals such that $r_1^{(3)}(k)p_{c,1} \geq r_2^{(3)}(k)p_{c,2} \geq \cdots \geq r_{N_0}^{(3)}(k)p_{c,N_0} > 0 \geq r_{N_0+1}^{(3)}(k)p_{c,N_0+1} \geq \cdots$
5: Transmit packets for clients 1 through N_0 by the ordering

---

Theorem 4.13 *The joint debt–channel policy is feasibility optimal.*

Proof. We first note that the joint debt-channel policy schedules an undelivered packet of client n that maximizes $r_n^{(3)}(k)p_{c,n}$ in every time slot.

Let $\hat{S}(t)$ be the collection of undelivered packets at the tth time slot of an interval, and $\hat{\mu}_n(\hat{S}(t), t)$ be the random variable that indicates the number of packets delivered for client n between the tth time slot and the end of the interval. To simplify notations, we represent each packet by the client that it is associated with. Let $V(\hat{S}(t), t)$ be the value of $E\{\sum_n r_n^{(3)}(k)^+ \hat{\mu}_n(\hat{S}(t), t)\}$ under the joint debt-channel policy. We show that $V(\hat{S}(t), t)$ is the maximum of $E\{\sum_n r_n^{(3)}(k)^+ \hat{\mu}_n(\hat{S}(t), t)\}$ among all policies, for all $t \in [1, T]$.

We prove this by induction on T. When $t = T$, the AP can make exactly one more transmission. If the AP schedules a packet for client m, $E\{\sum_n r_n^{(3)}(k)^+ \hat{\mu}_n(\hat{S}(t), t)\} = r_m^{(3)}(k)^+ p_{c,m}$, which is maximized by the joint debt-channel policy.

Assume that $V(\hat{S}(t), t)$ is the maximum of $E\{\sum_n r_n^{(3)}(k)^+ \hat{\mu}_n(\hat{S}(t), t)\}$ among all policies, for all $t \in [\hat{t} + 1, T]$. Now we consider the case when $t = \hat{t}$. Let m be the client that has an undelivered packet and maximizes $r_m^{(3)}(k)^+ p_{c,m}$, which is to be scheduled by the joint debt-channel policy. Consider some policy η and let $V^\eta(\hat{S}(t), t)$ be the value of $E\{\sum_n r_n^{(3)}(k)^+ \hat{\mu}_n(\hat{S}(t), t)\}$ under η.

Suppose η also schedules client m in the tth time slot. We then have $\hat{S}(t + 1) = \hat{S}(t)\backslash\{m\}$ with probability $p_{c,m}$, and $\hat{S}(t + 1) = \hat{S}(t)$ with probability $1 - p_{c,m}$, under both η and the joint debt-channel policy. Thus, we have

$$V^\eta(\hat{S}(t), t) = r_m^{(3)}(k)^+ p_{c,m} + p_{c,m} V^\eta(\hat{S}(t)\backslash\{m\}, t + 1) + (1 - p_{c,m}) V^\eta(\hat{S}(t), t + 1), \quad (4.18)$$

and

$$V(\hat{S}(t), t) = r_m^{(3)}(k)^+ p_{c,m} + p_{c,m} V(\hat{S}(t)\backslash\{m\}, t + 1) + (1 - p_{c,m}) V(\hat{S}(t), t + 1). \quad (4.19)$$

By the induction hypothesis, $V^\eta(\hat{S}(t)\backslash\{m\}, t + 1) \leq V(\hat{S}(t)\backslash\{m\}, t + 1)$ and $V^\eta(\hat{S}(t), t + 1) \leq V(\hat{S}(t), t + 1)$. Therefore, $V^\eta(\hat{S}(t), t) \leq V(\hat{S}(t), t)$.

On the other hand, suppose η schedules some client $l \neq m$ in the tth time slot. We can modify η so that it schedules l in the tth time slot, and then uses the joint debt-channel policy starting from the $(t + 1)^{th}$ time slot. By the induction hypothesis, this modification does not decrease $V^\eta(\hat{S}(t), t)$. Also, as client m maximizes $p_{c,m} r_m^{(3)}(k)$, the modified η schedules client m for transmission in the $(t + 1)^{th}$ time slot. The resulting policy is equivalent to one that schedules m in the tth time slot, l in the $(t + 1)^{th}$ time slot, and then uses the joint debt-channel policy starting from the $(t + 2)^{th}$ time slot. Hence, by the arguments in previous paragraphs, $V^\eta(\hat{S}(t), t) \leq V(\hat{S}(t), t)$.

In sum, $V(\hat{S}(t), t)$ is the maximum of $E\{\sum_n r_n^{(3)}(k)^+ \hat{\mu}_n(\hat{S}(t), t)\}$ among all policies, for $t \in [\hat{t}, T]$. By induction and Theorem 4.11, the joint debt-channel policy is feasibility optimal. $\square$

4.4   SCHEDULING POLICY UNDER RATE ADAPTATION

We now propose a feasibility optimal scheduling policy when rate adaptation is employed. We allow for the channel qualities to be time-varying, and for the clients to have differing deadlines.

To derive the scheduling policy, we use the delivery debt $r_n^{(3)}(k)$. Thus, $z_n(q_n) := q_n$, while $\mu_n(k)$ is the number of packets delivered for client n.

Suppose that at the beginning of interval k, the delivery debt vector is $[r_n^{(3)}(k)]$, the channel state is c, and the collection of arrived packets is S. If a client generates more than one packet, say, i packets, in the interval, we instead assume that there are i clients that correspond to this client in the system, with each of them corresponding to one packet that is generated by the original client. The i clients have the same delivery debt and channel condition as that of the original client. Hence, we can assume that each client generates at most one packet in an interval for the remainder of this section. This assumption only serves to simplify notation and is not restrictive.

We represent each packet in S by the client that generates it. The transmission time for client n is $s_{c,n}$ time slots, and client n has a delay bound of τ_n. Since transmissions are assumed to be error-free when rate adaptation is applied, the scheduling policy consists of finding an ordered subset $S' = \{m_1, m_2, \ldots, m_{N'}\}$ of S such that $\sum_{n=1}^{l} s_{c,m_n} \leq \tau_{m_l}$, for all $1 \leq l \leq m_{N'}$. That is, when clients are scheduled according to the ordering, no packets for clients in S' miss their respective delay bounds. By Theorem 4.11, a policy using an ordered set S' that maximizes $\sum_{n \in S'} r_n^{(3)}(k)$ with the above constraint is feasibility optimal. This is a variation of the knapsack problem. When S' is selected, reordering clients in S' in an earliest-deadline-first fashion also allows all packets to meet their respective delay bounds. Based on this observation, we derive the feasibility optimal scheduling algorithm, the Modified Knapsack Algorithm. Let $M[n, t]$ be the maximum debt a policy can collect if only clients 1 through n can be scheduled and all transmissions need to complete before time slot t. Thus, $\max_{S'} \sum_{n \in S'} r_n^{(3)}(k) = M[N, T]$. Also, iteratively:

$$M[n, t] = \begin{cases} M[n, t-1] & \text{if } t > \tau_n, \\ \max\{M[n-1, t], \\ \quad r_n^{(3)}(k) + M[n-1, t - s_{c,n}]\} & \text{otherwise,} \end{cases} \tag{4.20}$$

where $M[n-1, t]$ is the maximum debt that can be collected when client n is not scheduled, and $r_n^{(3)}(k) + M[n-1, t - s_{c,n}]$ is that when client n is scheduled. The complexity of this algorithm is $O(NT)$, and it is thus reasonably efficient.

Algorithm 3 Modified Knapsack Policy

1: **for** $n = 1$ to N **do**
2:      $r_n^{(3)}(k) \leftarrow q_n k - d_n(kT)$
3: **end for**
4: Sort clients such that $\tau_1 \leq \tau_2 \leq \cdots \leq \tau_N$
5: **for** $n = 1$ to N **do**
6:      $M[n, 0] \leftarrow 0$
7:      $S'[n, 0] \leftarrow \phi$
8: **end for**
9: **for** $n = 1$ to N **do**
10:     **for** $t = 1$ to T **do**
11:         **if** $t > \tau_n$ **then**
12:             $M[n, t] \leftarrow M[n, t - 1]$
13:             $S'[n, t] \leftarrow S'[n, t - 1]$
14:         **else if** client n has a packet AND
                $r_n^{(3)}(k) + M[n - 1, t - s_{c,n}] > M[n - 1, t]$ **then**
15:             $M[n, t] \leftarrow r_n^{(3)}(k) + M[n - 1, t - s_{c,n}]$
16:             $S'[n, t] \leftarrow S'[n - 1, t - s_{c,n}] + \{n\}$
17:         **else**
18:             $M[n, t] \leftarrow M[n - 1, t]$
19:             $S'[n, t] \leftarrow S'[n - 1, t]$
20:         **end if**
21:     **end for**
22: **end for**
23: schedule according to $S'[N, T]$

CHAPTER 5

Utility Maximization without Rate Adaptation

5.1   PROBLEM FORMULATION AND DECOMPOSITION

We now consider the case where the deadlines are fixed (i.e., inelastic), but the timely-throughput is elastic. In the previous chapters, it is assumed that the timely-throughput requirements, $[q_n]$, are given and fixed. In this chapter, we address the problem of how to choose $q := [q_n]$ so that the total utility of all the clients in the system can be maximized. We consider the scenario as discussed in Chapter 2.1, where rate adaptation is not applied, channel state is static, and all clients require the same delay bound T. As in Chapter 2, we use p_n to denote the channel reliability for client n.

We begin by supposing that each client has a certain utility function, $U_n(q_n)$, which is strictly increasing, strictly concave, and continuously differentiable function over the range $0 < q_n \leq 1$, with the value at 0 defined as the right limit, possibly $-\infty$. The problem of choosing q_n to maximize the total utility, under the feasibility constraint of Theorem 2.23, can be described by the following convex optimization problem:

SYSTEM:

$$\text{Max} \sum_{i=1}^{N} U_i(q_i) \tag{5.1}$$

$$\text{s.t.} \sum_{i \in S} \frac{q_i}{p_i} \leq T - I_S, \forall S \subseteq \{1, 2, \ldots, N\}, \tag{5.2}$$

$$\text{over } q_n \geq 0, \forall 1 \leq n \leq N. \tag{5.3}$$

The *SYSTEM* problem is a convex optimization problem, and we can apply standard techniques to solve it. However, solving *SYSTEM* directly can be difficult due to two major challenges. First, the AP may not know the utility function of each client. Second, the feasibility constraint (5.2) involves testing $\sum_{i \in S} \frac{q_i}{p_i} \leq T - I_S$ for every subset S. Hence, there are 2^N tests. To overcome these challenges, we decompose *SYSTEM* into two simpler problems, namely, *CLIENT* and *ACCESS-POINT*, as described below. This decomposition was earlier considered by Kelly [25], though in the context of dealing with rate control for non-real-time traffic.

Suppose client n is willing to pay an amount of ρ_n per interval, and receives a long-term average timely-throughput q_n proportional to ρ_n, with $\rho_n = \psi_n q_n$. That is, ψ_n is the "price" of one unit of timely-throughput. If $\psi_n > 0$, the utility maximization problem for client n is:

CLIENT$_n$:

$$\text{Max } U_n(\frac{\rho_n}{\psi_n}) - \rho_n \tag{5.4}$$

$$\text{over } 0 \le \rho_n \le \psi_n. \tag{5.5}$$

On the other hand, given that client n is willing to pay ρ_n per interval, we suppose that the AP wishes to find the vector q to maximize $\sum_{i=1}^{N} \rho_i \log q_i$, under the feasibility constraints. In other words, the AP has to solve the following optimization problem:

ACCESS-POINT:

$$\text{Max } \sum_{i=1}^{N} \rho_i \log q_i \tag{5.6}$$

$$\text{s.t. } \sum_{i \in S} \frac{q_i}{p_i} \le T - I_S, \forall S \subseteq \{1, 2, \ldots, N\}, \tag{5.7}$$

$$\text{over } q_n \ge 0, \forall 1 \le n \le N. \tag{5.8}$$

We begin by showing that solving $SYSTEM$ is equivalent to jointly solving $CLIENT_n$ and $ACCESS\text{-}POINT$.

Theorem 5.1   *There exist non-negative vectors q, $\rho := [\rho_n]$, and $\psi := [\psi_n]$, with $\rho_n = \psi_n q_n$, such that:*

(i) For n such that $\psi_n > 0$, ρ_n is a solution to $CLIENT_n$;

(ii) Given that each client n pays ρ_n per interval, q is a solution to $ACCESS\text{-}POINT$.

Further, if q, ρ, and ψ are all positive vectors, the vector q is also a solution to $SYSTEM$.

Proof. We will first show the existence of q, ρ, and ψ that satisfy (i) and (ii). We will then show that the resulting q is also the solution to $SYSTEM$.

There exists some $\epsilon > 0$ such that by letting $q_n \equiv \epsilon$, the vector q is an interior point of the feasible region for both $SYSTEM$ (5.2) (5.3), and $ACCESS\text{-}POINT$ (5.7) (5.8). Also, by setting $\rho_n \equiv \epsilon$, ρ_n is also an interior point of the feasible region for $CLIENT_n$ (5.5). Therefore, by Slater's condition [68], a feasible point for $SYSTEM$, $CLIENT_n$, or $ACCESS\text{-}POINT$, is the optimal solution for the respective problem if and only if it satisfies the corresponding Karush-Kuhn-Tucker (KKT) condition [68] for the problem. Further, since the feasible region for each of the problems is compact, and the utilities are continuous on it, or since the utility converges to $-\infty$ at $q_n = 0$, there exists an optimal solution to each of them.

The Lagrangian of $SYSTEM$ is:

$$L_{SYS}(q, \lambda, \nu) := -\sum_{i=1}^{N} U_i(q_i) + \sum_{S \subseteq \{1,2,...,N\}} \lambda_S [\sum_{i \in S} \frac{q_i}{p_i} - (T - I_S)] - \sum_{i=1}^{N} \nu_i q_i, \qquad (5.9)$$

where $\lambda := [\lambda_S : S \subseteq \{1, 2, \ldots, N\}]$ and $\nu := [\nu_n : 1 \le n \le N]$ are the Lagrange multipliers. By the KKT condition, a vector $q^* := [q_1^*, q_2^*, \ldots, q_N^*]$ is the optimal solution to SYSTEM if q^* is feasible and there exists vectors λ^* and ν^* such that:

$$\frac{\partial L_{SYS}}{\partial q_n}\bigg|_{q^*,\lambda^*,\nu^*} = -U_n'(q_n^*) + \frac{\sum_{S \ni n} \lambda_S^*}{p_n} - \nu_n^*$$

$$= 0, \forall 1 \le n \le N, \qquad (5.10)$$

$$\lambda_S^* [\sum_{i \in S} \frac{q_i^*}{p_i} - (T - I_S)] = 0, \forall S \subseteq \{1, 2, \ldots, N\}, \qquad (5.11)$$

$$\nu_n^* q_n^* = 0, \forall 1 \le n \le N, \qquad (5.12)$$

$$\lambda_S^* \ge 0, \forall S \subseteq \{1, \ldots, N\}, \text{ and } \nu_n^* \ge 0, \forall 1 \le n \le N. \qquad (5.13)$$

The Lagrangian of $CLIENT_n$ is:

$$L_{CLI}(\rho_n, \xi_n) := -U_n(\frac{\rho_n}{\psi_n}) + \rho_n - \xi_n \rho_n,$$

where ξ_n is the Lagrange multiplier for $CLIENT_n$. By the KKT condition, ρ_n^* is the optimal solution to $CLIENT_n$ if $\rho_n^* \ge 0$ and there exists ξ_n^* such that:

$$\frac{dL_{CLI}}{d\rho_n}\bigg|_{\rho_n^*,\xi_n^*} = -\frac{1}{\psi_n} U_n'(\frac{\rho_n^*}{\psi_n}) + 1 - \xi_n^* = 0, \qquad (5.14)$$

$$\xi_n^* \rho_n^* = 0, \qquad (5.15)$$

$$\xi_n^* \ge 0. \qquad (5.16)$$

Finally, the Lagrangian of $ACCESS\text{-}POINT$ is:

$$L_{NET}(q, \zeta, \mu) := -\sum_{i=1}^{N} \rho_i \log q_i + \sum_{S \subseteq \{1,2,...,N\}} \zeta_S [\sum_{i \in S} \frac{q_i}{p_i} - (T - I_S)] - \sum_{i=1}^{N} \mu_i q_i, \qquad (5.17)$$

where $\zeta := [\zeta_S : S \subseteq \{1, 2, \ldots, N\}]$ and $\mu := [\mu_n : 1 \le n \le N]$ are the Lagrange multipliers. Again, by the KKT condition, a vector $q^* := [q_n^*]$ is the optimal solution to $ACCESS\text{-}POINT$ if

q^* is feasible and there exist vectors ζ^* and μ^* such that:

$$\left.\frac{\partial L_{NET}}{\partial q_n}\right|_{q^*,\zeta^*,\mu^*} = -\frac{\rho_n}{q_n^*} + \frac{\sum_{S \ni n} \zeta_S^*}{p_n} - \mu_n^*$$
$$= 0, \forall 1 \le n \le N, \tag{5.18}$$

$$\zeta_S^*\left[\sum_{i \in S} \frac{q_i^*}{p_i} - (T - I_S)\right] = 0, \forall S \subseteq \{1, 2, \ldots, N\}, \tag{5.19}$$

$$\mu_n^* q_n^* = 0, \forall 1 \le n \le N, \tag{5.20}$$
$$\zeta_S^* \ge 0, \forall S \subseteq \{1, \ldots, N\}, \text{ and } \mu_n^* \ge 0, \forall 1 \le n \le N. \tag{5.21}$$

Let q^* be a solution to $SYSTEM$, and let λ^*, ν^* be the corresponding Lagrange multipliers that satisfy conditions (5.10)–(5.13). Let $q_n = q_n^*$, $\rho_n = \frac{\sum_{S \ni n} \lambda_S^*}{p_n} q_n^*$, and $\psi_n = \frac{\sum_{S \ni n} \lambda_S^*}{p_n}$, for all n. Clearly, q, ρ, and ψ are all non-negative vectors. We will show (q, ρ, ψ) satisfy (i) and (ii).

We first show (i) for all n such that $\psi_n = \frac{\sum_{S \ni n} \lambda_S^*}{p_n} > 0$. It is obvious that $\rho_n = \psi_n q_n$. Also, $\rho_n \ge 0$, since $\lambda_S^* \ge 0$ (by (5.13)) and $q_n^* \ge 0$ (since q^* is feasible). Further, let the Lagrange multiplier of $CLIENT_n$, ξ_n, be equal to $\nu_n^* / \frac{\sum_{S \ni n} \lambda_S^*}{p_n} = \nu_n^*/\psi_n$. Then we have:

$$\left.\frac{\partial L_{CLI}}{\partial \rho_n}\right|_{\rho_n,\xi_n} = -\frac{1}{\psi_n} U_n'\left(\frac{\rho_n}{\psi_n}\right) + 1 - \xi_n$$
$$= \frac{1}{\psi_n}\left(-U_n'\left(\frac{\rho_n}{\psi_n}\right) + \psi_n - \psi_n \xi_n\right)$$
$$= \frac{1}{\psi_n}\left(-U_n'(q_n^*) + \frac{\sum_{S \ni n} \lambda_S^*}{p_n} - \nu_n^*\right) = 0, \text{ by (5.10)},$$

$$\xi_n \rho_n = \frac{\nu_n^*}{\psi_n} \psi_n q_n^* = \nu_n^* q_n^* = 0, \text{ by (5.12)}$$

$$\xi_n = \nu_n^* / \frac{\sum_{S \ni n} \lambda_S^*}{p_n} \ge 0, \text{ by (5.13)}.$$

In sum, (ρ, ψ, ξ) satisfies the KKT conditions for $CLIENT_n$, and therefore ρ_n is a solution to $CLIENT_n$, with $\rho_n = \psi_n q_n$.

Next we establish (ii). Since $q = q^*$ is the solution to $SYSTEM$, it is feasible. Let the Lagrange multipliers of $ACCESS-POINT$ be $\zeta_S = \lambda_S^*$, $\forall S$, and $\mu_n = 0$, $\forall n$, respectively. Given that each client n pays ρ_n per interval, we have:

$$\left.\frac{\partial L_{NET}}{\partial q_n}\right|_{q,\zeta,\mu} = -\frac{\rho_n}{q_n} + \frac{\sum_{S \ni n} \zeta_S}{p_n} - \mu_n$$
$$= -\psi_n + \psi_n - 0 = 0, \forall n,$$

$$\zeta_S\left[\sum_{i \in S} \frac{q_i}{p_i} - (T - I_S)\right] = \lambda_S^*\left[\sum_{i \in S} \frac{q_i^*}{p_i} - (T - I_S)\right]$$
$$= 0, \forall S, \text{ by (5.11)},$$

$$\mu_n q_n = 0 \times q_n = 0, \forall n,$$
$$\zeta_S = \lambda_S^* \ge 0, \forall S \text{ (by (5.13))}, \text{ and } \mu_n \ge 0, \forall n.$$

Therefore, (q, ζ, μ) satisfies the KKT condition for $ACCESS\text{-}POINT$ and thus q is a solution to $ACCESS\text{-}POINT$.

For the converse, suppose (q, ρ, ψ) are positive vectors with $\rho_n = \psi_n q_n$, for all n, that satisfy (i) and (ii). We wish to show that q is a solution to $SYSTEM$. Let ξ_n be the Lagrange multiplier for $CLIENT_n$. Since we assume $\psi_n > 0$ for all n, the problem $CLIENT_n$ is well-defined for all n, and so is ξ_n. Also, let ζ and μ be the Lagrange multipliers for $ACCESS\text{-}POINT$. Since $q_n > 0$ for all n, we have $\mu_n = 0$ for all n by (5.20). By (5.18), we also have:

$$\left.\frac{\partial L_{NET}}{\partial q_n}\right|_{q,\zeta,\mu} = -\frac{\rho_n}{q_n} + \frac{\sum_{S \ni n} \zeta_S}{p_n} - \mu_n$$

$$= -\psi_n + \frac{\sum_{S \ni n} \zeta_S}{p_n} = 0,$$

and thus $\psi_n = \frac{\sum_{S \ni n} \zeta_S}{p_n}$. Let $\lambda_S := \zeta_S$ for all S, and $\nu_n := \psi_n \xi_n$ for all n. We claim that q is the optimal solution to $SYSTEM$ with Lagrange multipliers λ and ν.

Since q is a solution to $ACCESS\text{-}POINT$, it is feasible. Further, we have:

$$\left.\frac{\partial L_{SYS}}{\partial q_n}\right|_{q,\lambda,\nu} = -U_n'(q_n) + \frac{\sum_{S \ni n} \lambda_S}{p_n} - \nu_n$$

$$= -U_n'\left(\frac{\rho_n}{\psi_n}\right) + \psi_n - \psi_n \xi_n = 0, \forall n, \text{ by (5.14)},$$

$$\lambda_S\left[\sum_{n \in S} \frac{q_n}{p_n} - (T - I_S)\right] = \zeta_S\left[\sum_{n \in S} \frac{q_n}{p_n} - (T - I_S)\right]$$

$$= 0, \forall S, \text{ by (5.19)},$$

$$\nu_n q_n = \xi_n \rho_n = 0, \forall n, \text{ by (5.15)},$$

$$\lambda_S = \zeta_S \geq 0, \forall S, \text{ by (5.21)},$$

$$\nu_n = \psi_n \xi_n \geq 0, \forall n, \text{ by (5.16)}.$$

Thus, (q, λ, ν) satisfy the KKT condition for $SYSTEM$, and so q is a solution to $SYSTEM$. $\qquad\square$

5.2   A BIDDING PROCEDURE BETWEEN CLIENTS AND ACCESS POINT

Above, we have shown that the maximum total utility of the system can be achieved when the solutions to the problems $CLIENT_n$ and $ACCESS\text{-}POINT$ agree. In this section, we formulate a bidding procedure for such reconciliation. We also discuss the meanings of the problems $CLIENT_n$ and $ACCESS\text{-}POINT$ in this bidding procedure.

The bidding procedure is as follows:

1: Each client n announces an amount ρ_n that it pays per interval.

2: After noting the amounts, $\rho_1, \rho_2, \ldots, \rho_N$, paid by the clients, the AP chooses a scheduling policy so that the resulting long-term timely throughput, q_n, for each client maximizes $\sum_{i=1}^{N} \rho_i \log q_i$, subject to feasibility of $[q_n]$.

3: The client n observes its own timely-throughput, q_n. It computes $\psi_n := \rho_n/q_n$. It then determines $\rho_n^* \geq 0$ to maximize $U_n(\frac{\rho_n^*}{\psi_n}) - \rho_n^*$. Client n then updates the amount it pays to $(1 - \alpha)\rho_n + \alpha\rho_n^*$, with some fixed $0 < \alpha < 1$, and announces the new bid value.

4: Go back to Step 2.

In Step 3 of the procedure, client n chooses its new amount of payment as a weighted average of the past amount and the derived optimal value, instead of the derived optimal value. This design serves two purposes. First, it seeks to avoid the system oscillating between two extreme values. Second, since ρ_n is initiated to a positive value, and ρ_n^* derived in each iteration is always non-negative, this design guarantees ρ_n to be positive throughout all iterations. Since $\psi_n := \rho_n/q_n$, this also ensures that $\psi_n > 0$ and that the function $U_n(\frac{\rho_n}{\psi_n})$ is consequently always well-defined.

We show that the fixed point of this bidding procedure maximizes the total utility of the system:

Theorem 5.2   *Suppose that at the fixed point of the bidding procedure, each client n pays ρ_n^* per interval, and receives timely-throughput q_n^*. If both ρ_n^* and q_n^* are positive for all n, then the vector q^* maximizes the total utility of the system.*

Proof. Let $\psi_n^* = \frac{\rho_n^*}{q_n^*}$. It is positive since both ρ_n^* and q_n^* are positive. Since the vectors q^* and ρ^* are derived from the fixed point, ρ_n^* maximizes $U_n(\frac{\rho_n}{\psi_n^*}) - \rho_n$, over all $\rho_n \geq 0$, as described in Step 3 of the procedure. Thus, ρ_n^* is a solution to $CLIENT_n$, given $\rho_n^* = \psi_n^* q_n^*$. Similarly, from Step 2, q^* is the feasible vector that maximizes $\sum_{i=1}^{N} \rho_i^* \log q_i$, over all feasible vectors q. Thus, q^* is a solution to $ACCESS\text{-}POINT$, given that each client n pays ρ_n^* per interval. By Theorem 5.1, q^* is the unique solution to $SYSTEM$ and therefore maximizes the total utility of the system. □

Next, we describe the meaning of the procedure. In Step 3, client n assumes a linear relation between the amount it pays, ρ_n, and the timely-throughput it receives, q_n. To be more precise, it assumes $\rho_n = \psi_n q_n$, where ψ_n is the price. Thus, maximizing $U_n(\frac{\rho_n}{\psi_n}) - \rho_n$ is equivalent to maximizing $U_n(q_n) - \rho_n$. Recall that $U_n(q_n)$ is the utility that client n obtains when it receives timely-throughput q_n. $U_n(q_n) - \rho_n$ is therefore the net profit that client n gets. In short, in Step 3, the goal of client n is to selfishly maximize its own net profit using a first order linear approximation to the relation between payment and timely-throughput.

We next discuss the behavior of the AP in Step 2. The AP schedules clients so that the resulting timely-throughput vector q is a solution to the problem $ACCESS\text{-}POINT$, given that each client n pays ρ_n per interval. Thus, q is feasible and there exist vectors ζ and μ that satisfy conditions (5.18)–(5.21). While it is difficult to solve this problem, we consider a special restrictive case that gives us a simple solution and insight into the AP's behavior, which we will subsequently generalize. Let $TOT := \{1, 2, \ldots, N\}$ be the set that consisting of all clients. We assume that a solution (q, ζ, μ) to the problem has the following properties: $\zeta_S = 0$, for all $S \neq TOT$, $\zeta_{TOT} > 0$, and $\mu_n = 0$, for all n. By (5.18), we have:

$$-\frac{\rho_n}{q_n} + \frac{\sum_{S \ni n} \zeta_S}{p_n} - \mu_n = -\frac{\rho_n}{q_n} + \frac{\zeta_{TOT}}{p_n} = 0,$$

and therefore $q_n = p_n \rho_n / \zeta_{TOT}$. Further, since $\zeta_{TOT} > 0$, (5.19) requires that:

$$\sum_{i \in TOT} \frac{q_i}{p_i} - (T - I_{TOT}) = \sum_{i \in TOT} \frac{\rho_i}{\zeta_{TOT}} - (T - I_{TOT}) = 0.$$

Thus, $\zeta_{TOT} = \frac{\sum_{i=1}^{N} \rho_i}{T - I_{TOT}}$ and $\frac{q_n}{p_n} = \frac{\rho_n}{\sum_{i=1}^{N} m_i}(T - I_{TOT})$, for all n. Notice that the derived (q, ζ, μ) satisfies conditions (5.18)–(5.21). Thus, under the assumption that q is feasible, this special case actually maximizes $\sum_{i=1}^{N} \rho_i \log q_i$. In Section 5.3 we will address the general situation without any such assumption, since it need not be true.

Recall that I_{TOT} is the average number of time slots that the AP is forced to be idle in an interval after it has completed all clients. Also, by Lemma 2.5, $\frac{q_n}{p_n}$ is the workload of client n, that is, the average number of time slots that the AP should spend working for client n. Thus, letting $\frac{q_n}{p_n} = \frac{\rho_n}{\sum_{i=1}^{N} \rho_i}(T - I_{TOT})$ for all n, the AP tries to allocate those non-idle time slots so that the average number of time slots each client gets is proportional to its payment. Although we only study the special case of I_{TOT} here, we will show that the same behavior also holds for the general case in Section 5.3.

In summary, the procedure proposed in this section actually describes a bidding procedure, where clients are bidding for non-idle time slots. Each client gets a share of time slots that is proportional to its bid. The AP thus assigns timely-throughputs, based on which the clients calculate a price and selfishly maximize their own net profits. Finally, Theorem 5.2 states that the equilibrium point of this procedure maximizes the total utility of the system.

5.3 A SCHEDULING POLICY FOR THE ACESS POINT

In Section 5.2, we have shown that by setting $q_n := p_n \frac{\rho_n}{\sum_{i=1}^{N} \rho_i}(T - I_{TOT})$, the resulting vector q solves *ACCESS-POINT* provided q is indeed feasible. Unfortunately, such q is not always feasible and solving *ACCESS-POINT* is in general difficult. Even for the special case discussed in Section 5.2, solving *ACCESS-POINT* requires knowledge of channel conditions, that is, p_n. In this section, we propose a very simple priority-based scheduling policy that can achieve the optimal solution for *ACCESS-POINT*, and that too without any knowledge of the channel conditions.

In the special case discussed in Section 5.2, the AP tries, though it may be impossible in general, to allocate non-idle time slots to clients in proportion to their payments. Based on this intuitive guideline, we design the following scheduling policy. Let $u_n(t)$ be the number of time slots that the AP has allocated for client n up to time t. At the beginning of each interval, the AP sorts all clients in increasing order of $\frac{u_n(t)}{\rho_n}$, so that $\frac{u_1(t)}{\rho_1} \leq \frac{u_2(t)}{\rho_2} \leq \ldots$ after renumbering clients if necessary.

The AP then schedules transmissions according to the priority ordering, where clients with smaller $\frac{u_n(t)}{\rho_n}$ get higher priorities. Specifically, in each time slot during the interval, the AP chooses the smallest i for which the packet for client i is not yet delivered, and then transmits the packet for client i in that time slot. We call this the *weighted transmission* policy (WT). Notice that the policy only requires the AP to keep track of the bids of clients and the number of time slots each client has been allocated in the past, followed by a sorting of $\frac{u_n(t)}{\rho_n}$ among all clients. Thus, the policy requires no information on the actual channel conditions, and is tractable. Simple as it is, we show that the policy actually achieves the optimal solution for $ACCESS\text{-}POINT$. In the following sections, we first prove that the vector of timely throughputs resulting from the WT policy converges to a single point. We then prove that this limit is the optimal solution for $ACCESS\text{-}POINT$.

5.3.1   CONVERGENCE OF THE WEIGHTED TRANSMISSION POLICY

We now prove that, by applying the WT policy, the timely-throughputs of clients will converge to a vector q. To do so, we actually prove the convergence property and precise limit of a more general class of scheduling policies, which not only consists of the WT policy but also the largest time-based debt scheduling policy considered earlier. The proof is similar to that used in Chapter 2 and is also based on Blackwell's approachability theorem [67].

Now we formulate our more general class of scheduling policies. We call a policy a *generalized transmission time policy* if, for a choice of a positive parameter vector a and non-negative parameter vector b, the AP sorts clients by $a_n u_n(t) - b_n t$ at the beginning of each interval, and gives priorities to clients with lower values of this quantity. Note that the special case $a_n \equiv \frac{1}{\rho_n}$ and $b_n \equiv 0$ yields the WT policy, while the choice $a_n \equiv 1$ and $b_n \equiv \frac{q_n}{Tp_n}$ yields the largest time-based debt first policy.

Theorem 5.3   *For each generalized transmission time policy, there exists a vector q such that the vector of workloads resulting from the policy converges to $w(q) := [w_n(q_n)]$.*

Proof.   Given the parameters $\{(a_n, b_n) : 1 \leq n \leq N\}$, we give an exact expression for the limiting q. We define a sequence of sets $\{H_k\}$ and corresponding values $\{\theta_k\}$ iteratively as follows. Let $H_0 := \phi$, $\theta_0 := -\infty$, and

$$H_k := \arg\min_{S : S \supsetneq H_{k-1}} \frac{\frac{1}{T}(I_{H_{k-1}} - I_S) - \sum_{n \in S \setminus H_{k-1}} \frac{b_n}{a_n}}{\sum_{n \in S \setminus H_{k-1}} 1/a_n},$$

$$\theta_k := \frac{\frac{1}{T}(I_{H_{k-1}} - I_{H_k}) - \sum_{n \in H_k \setminus H_{k-1}} \frac{b_n}{a_n}}{\sum_{n \in H_k \setminus H_{k-1}} 1/a_n}, \text{ for all } k > 0.$$

In selecting H_k, we always choose a maximal subset, breaking ties arbitrarily. $(H_1, \theta_1), (H_2, \theta_2), \ldots,$ can be iteratively defined until every client is in some H_k. Also, by the definition, we have $\theta_k > \theta_{k-1}$, for all $k > 0$. If client n is in $H_k \setminus H_{k-1}$, we define $q_n := Tp_n \frac{b_n + \theta_k}{a_n}$, and so $w_n(q_n) = T \frac{b_n + \theta_k}{a_n}$. The proof of convergence consists of two parts. First we prove that the vector of *work performed*, defined as the vector of average numbers of time slots that the AP spends on transmitting the packet for each client, approaches the set $\{w^*|w_n^* \geq w_n(q_n)\}$. Then we prove that $w(q)$ is the only feasible vector in the set $\{w^*|w_n^* \geq w_n(q_n)\}$. Since the *feasible region for workloads*, defined as the set of all feasible vectors for workloads, is approachable under any policy, the vector of work performed resulting from the generalized transmission time policy must converge to $w(q)$.

For the first part, we prove the following statement: for each $k \geq 1$, the set $W_k := \{w^*|w_n^* \geq T\frac{b_n+\theta_k}{a_n}, \forall n \notin H_{k-1}\}$ is approachable. Since $\cap_{i\geq 0}W_i = \{w^*|w_n^* \geq w_n(q_n)\}$, we also thereby prove that $\{w^*|w_n^* \geq w_n(q_n)\}$ is approachable.

Consider a linear transformation on the space of workloads $L(w) := [l_n : l_n = \frac{a_n w_n/T - b_n}{\sqrt{a_n}}]$. Proving W_k is approachable is equivalent to proving that its image under L, $V_k := \{l|l_n \geq \frac{\theta_k}{\sqrt{a_n}}, \forall n \notin H_{k-1}\}$, is approachable. Now we apply Blackwell's Theorem. Suppose that at some time t that is the beginning of an interval, the number of time slots that the AP has worked on client n is $u_n(t)$. The work performed for client n is $\frac{u_n(t)}{t/T}$, and the image of the vector of work performed under L is $x(t) := [x_n(t)|x_n(t) = \frac{a_n u_n(t)/t - b_n}{\sqrt{a_n}}]$, which we shall suppose is not in V_k. The generalized transmission time policy sorts clients so that $a_1 u_1(t) - b_1 \leq a_2 u_2(t) - b_2 \leq \ldots$, or equivalently, $\sqrt{a_1}x_1(t) \leq \sqrt{a_2}x_2(t) \leq \ldots$. The closest point in V_k to $x(t)$ is $y := [y_n]$, where $y_n = \frac{\theta_k}{\sqrt{a_n}}$, if $x_n(t) < \frac{\theta_k}{\sqrt{a_n}}$ and $n \notin H_{k-1}$, and $y_n = x_n$, otherwise. The hyperplane that passes through y and is orthogonal to the line segment xy is:

$$\{z|f(z) := \sum_{n:n\leq n_0, n\notin H_{k-1}} (z_n - \frac{\theta_k}{\sqrt{a_n}})(x_n(t) - \frac{\theta_k}{\sqrt{a_n}}) = 0\}.$$

Let π_n be the expected number of time slots that the AP spends on working for client n in this interval under the generalized transmission time policy. The image under L of the expected reward in this interval is $\pi_L := [\frac{a_n \pi_n/T - b_n}{\sqrt{a_n}}]$. Blackwell's Theorem shows that V_k is approachable if $x(t)$ and π_L are separated by the plane $\{z|f(z) = 0\}$. Since $f(x(t)) \geq 0$, it suffices to show that $f(\pi_L) \leq 0$.

We manipulate the original ordering, for this interval, so that all clients in H_{k-1} have higher priorities than those not in H_{k-1}, while preserving the relative ordering between clients not in H_{k-1}. Note that this manipulation will not give any client $n \notin H_{k-1}$ higher priority than it had in the original ordering. Therefore, π_n will not increase for any $n \notin H_{k-1}$. Since the value of $f(\pi_L)$ only depends on π_n for $n \notin H_{k-1}$, and increases as those π_n decrease, this manipulation will not decrease the value of $f(\pi_L)$. Thus, it suffices to prove that $f(\pi_L) \leq 0$, under this new ordering. Let $n_0 := |H_{k-1}| + 1$. Under this new ordering, we have: $\sqrt{a_{n_0}}x_{n_0}(t) \leq \sqrt{a_{n_0+1}}x_{n_0+1}(t) \leq \cdots \leq \sqrt{a_{n_1}}x_{n_1}(t) < \theta_k \leq \sqrt{a_{n_1+1}}x_{n_1+1}(t) \leq \cdots$.

Let $\delta_n = \sqrt{a_n}x_n(t) - \sqrt{a_{n+1}}x_{n+1}(t)$, for $n_0 \leq n \leq n_1 - 1$ and $\delta_{n_1} = \sqrt{a_{n_1}}x_{n_1}(t) - \theta_k$. Clearly, $\delta_n \leq 0$, for all $n_0 \leq n \leq n_1$. Now we can derive:

$$f(\pi_L) = \sum_{n=n_0}^{n_1}(\frac{a_n\pi_n/T - b_n}{\sqrt{a_n}} - \frac{\theta_k}{\sqrt{a_n}})(x_n(t) - \frac{\theta_k}{\sqrt{a_n}})$$

$$= \sum_{n=n_0}^{n_1}(\frac{\pi_n}{T} - \frac{b_n}{a_n} - \frac{\theta_k}{a_n})(\sqrt{a_n}x_n(t) - \theta_k)$$

$$= \sum_{i=n_0}^{n_1}(\frac{\sum_{n=n_0}^{i}\pi_n}{T} - \sum_{n=n_0}^{i}\frac{b_n}{a_n} - \theta_k\sum_{n=n_0}^{i}\frac{1}{a_n})\delta_i.$$

Recall that I_S is the expected number of idle time slots when the AP only caters to the subset S. Thus, under this ordering, we have $\sum_{n=1}^{i}\pi_n = T - I_{\{1,...,i\}}$, for all i, and $\sum_{n=n_0}^{i}\pi_n = I_{\{1,...,n_0-1\}} - I_{\{1,...,i\}} = I_{H_{k-1}} - I_{\{1,...,i\}}$, for all $i \geq n_0$. By the definition of H_k and θ_k, we also have

$$\frac{\sum_{n=n_0}^{i}\pi_n}{T} - \sum_{n=n_0}^{i}\frac{b_n}{a_n} - \theta_k\sum_{n=n_0}^{i}\frac{1}{a_n}$$

$$= (\sum_{n=n_0}^{i}\frac{1}{a_n})(\frac{\frac{1}{T}(I_{H_{k-1}} - I_{\{1,...,i\}}) - \sum_{n=n_0}^{i}\frac{b_n}{a_n}}{\sum_{n\in\{1,...,i\}\setminus H_{k-1}}1/a_n} - \theta_k) \geq 0.$$

Therefore, $f(\pi_L) \leq 0$, since $\delta_i \leq 0$, and V_k is indeed approachable, for all k.

We have established that the set $\{w^*|w_n^* \geq w_n(q_n)\}$ is approachable. Next we prove that $[w_n(q_n)]$ is the only feasible vector in the set. Consider any vector $w' \neq w(q)$ in the set. We have $w_n' \geq w_n(q_n)$ for all n, and $w_{n_0}' > w_{n_0}(q_{n_0})$, for some n_0. Suppose $n_0 \in H_k\setminus H_{k-1}$. We have:

$$\sum_{n\in H_k}w_n' > \sum_{n\in H_k}w_n(q_n) = \sum_{i=1}^{k}\sum_{n\in H_i\setminus H_{i-1}}T\frac{b_n + \theta_k}{a_n}$$

$$= \sum_{i=1}^{k}(I_{H_{i-1}} - I_{H_i}) = T - I_{H_k},$$

and thus w' is not feasible. Therefore, $w(q)$ is the only feasible vector in $\{w^*|w_n^* \geq w_n(q_n)\}$, and the vector of work performed resulting from the generalized transmission time policy must converge to $w(q)$.   □

Corollary 5.4   *For the policy of Theorem 5.3, the vector of timely-throughputs converges to q.*

Proof. Follows from Lemma 2.5.   □

5.3.2   OPTIMALITY OF THE WEIGHTED TRANSMISSION POLICY

We have shown that the vector of timely-throughputs of clients converge to a vector q under the WT policy. We now prove that this vector q is indeed the optimal solution to the *ACCESS-POINT* problem.

Theorem 5.5   *Given $[\rho_n]$, the vector q of long-term average timely throughputs resulting from the WT policy is a solution to ACCESS-POINT.*

Proof. We use the sequence of sets $\{H_k\}$ and values $\{\theta_k\}$, with $a_n := \frac{1}{\rho_n}$ and $b_n := 0$, as defined in the proof of Theorem 5.3. Let $K := |\{\theta_k\}|$. Thus, we have $H_K = TOT = \{1, 2, \ldots, N\}$. Also, let $m_k := |H_k|$. For convenience, we renumber clients so that $H_k = \{1, 2, \ldots, m_k\}$. The proof of Theorem 5.3 shows that $q_n = Tp_n\theta_k\rho_n$, for $n \in H_k \setminus H_{k-1}$. Therefore, $w_n(q_n) = \frac{q_n}{p_n} = T\theta_k\rho_n$. Obviously, q is feasible, since it is indeed achieved by the WT policy. Thus, to establish optimality, we only need to prove the existence of vectors ζ and μ that satisfy conditions (5.18)–(5.21).

Set $\mu_n := 0$ for all n. Let $\zeta_{H_K} = \zeta_{TOT} := \frac{\rho_N}{w_N(q_N)} = \frac{1}{T\theta_K}$ and $\zeta_{H_k} := \frac{\rho_{m_k}}{w_{m_k}(q_{m_k})} - \frac{\rho_{m_{k+1}}}{w_{m_{k+1}}(q_{m_{k+1}})} = \frac{1}{T\theta_k} - \frac{1}{T\theta_{k+1}}$, for $1 \leq k \leq K - 1$. Finally, let $\zeta_S := 0$, for all $S \notin \{H_1, H_2, \ldots, H_K\}$. We claim that the vectors ζ and μ, along with q, satisfy conditions (5.18)–(5.21).

We first evaluate condition (5.18). Suppose client n is in $H_k \setminus H_{k-1}$. Then client n is also in $H_{k+1}, H_{k+2}, \ldots, H_K$. So,

$$-\frac{\rho_n}{q_n} + \frac{\sum_{S \ni n} \zeta_S}{p_n} - \mu_n = -\frac{1}{T\theta_k p_n} + \frac{\sum_{i=k}^{K} \zeta_{H_i}}{p_n}$$

$$= -\frac{1}{T\theta_k p_n} + \frac{1}{T\theta_k p_n} = 0.$$

Thus, condition (5.18) is satisfied.

Since $\mu_n = 0$, for all n, condition (5.20) is satisfied. Further, since $\frac{1}{\theta_k} > \frac{1}{\theta_{k+1}}$, for all $1 \leq k \leq K - 1$, condition (5.21) is also satisfied. It remains to establish condition (5.19). Since $\zeta_S = 0$ for all $S \notin \{H_1, H_2, \ldots, H_K\}$, we only need to show $\sum_{i \in S} \frac{q_i}{p_i} - (T - I_S) = 0$ for $S \in \{H_1, H_2, \ldots, H_K\}$.

Consider H_k. For each client $i \in H_k$ and each client $j \notin H_k$, $\frac{w_i(q_i)}{\rho_i} < \frac{w_j(q_j)}{\rho_j}$. Since $w_n(q_n)$ is the average number of time slots that the AP spends on working for client n, we have $\frac{u_i(t)}{\rho_i} < \frac{u_j(t)}{\rho_j}$, for all $i \in H_k$ and $j \notin H_k$, after a finite number of intervals. Therefore, except for a finite number of intervals, clients in H_k will have priority over those not in H_k. In other words, if we only consider the behavior of those clients in H_k, it is the same as if the AP only works on the subset H_k of clients. Further, recall that I_{H_k} is the expected number of time slots that the AP is forced to stay idle when the AP only works on the subset H_k of clients. Thus, we have $\sum_{i \in H_k} w_i(q_i) = T - I_{H_k}$ and $\sum_{i \in H_k} \frac{q_i}{p_i} - (T - I_{H_k}) = 0$, for all k. $\qquad\square$

5.4 SIMULATION RESULTS

In this section, we present the simulation results of the total utility that is achieved by iterating between the bidding procedure and the WT policy, which we call WT-Bid. We assume that the utility function of each client n is given by $\gamma_n \frac{q_n^{\alpha_n}-1}{\alpha_n}$, where γ_n is a positive integer and $0 < \alpha_n < 1$. This utility function is strictly increasing, strictly concave, and differentiable for any γ_n and α_n. In addition to evaluating the policy WT-Bid, we also compare the results of three other policies: a policy that employs the WT policy but without updating the bids from clients, which we call WT-NoBid; a policy that decides priorities randomly among clients at the beginning of each interval, which we call Rand; and a policy that gives clients with larger γ_n higher priorities, with ties broken randomly, which we call P-Rand.

In each simulation, we assume there are 30 clients. The nth client has channel reliability $p_n = (50 + n)\%$, $\gamma_n = (n \mod 3) + 1$, and $\alpha_n = 0.3 + 0.1(n \mod 5)$. In addition to plotting the average of total utility over all simulation runs, we also plot the variance of total utility.

We present the simulation results for VoIP traffic and use the same simulation set up as in Chapter 2.5. Figures 5.1 and 5.2 show the simulation results. The WT-Bid policy not only achieves

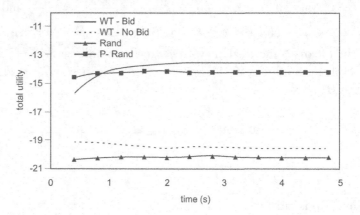

Figure 5.1: Average of total utility

the highest average total utility but also the smallest variance. This result suggests that the WT-Bid policy converges quickly. On the other hand, the WT-NoBid policy fails to provide satisfactory performance since it does not consider the different utility functions that clients may have. The P-Rand policy offers much better performance than both the WT-NoBid policy and the Rand policy since it correctly gives higher priority to clients with higher γ_n. Still, it cannot differentiate between clients with the same γ_n and thus can only provide suboptimal performance.

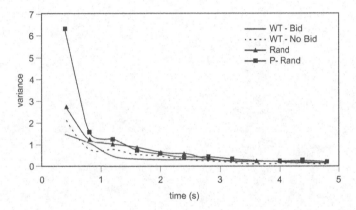

Figure 5.2: Variance of total utility

CHAPTER 6

Utility Maximization with Rate Adaptation

In this chapter, we treat the problem of utility maximization when rate adaptation is applied. We introduce a more general model that can be applied to not only delay-constrained wireless networks but also a variety of other applications. In addition, we address the selfish behaviors of clients. We employ an on-line auction design that achieves the maximum total utility in the network as well as preventing selfish clients from gaining additional net utility by lying about their utility function. Hence, the auction design is both optimal and incentive compatible.

6.1 PROBLEM OVERVIEW

Consider a wireless system with one server and N clients, numbered $\{1, 2, \ldots, N\}$. Time is divided into *time intervals*. Each client desires some service in each time interval. The service requirement within a time interval of each client is indivisible; that is, the server can only either fully meet the demand of a client or not serve it at all. At the beginning of each time interval, the server obtains the current channel condition. Both the demands of clients and the channel condition can be time-varying, and together we call them the *system state* in each time interval. The sever can learn the system state by either polling, probing, or estimating. Since these operations are costly and cannot be carried out too frequently, the server assumes that the system state does not change within an interval. Due to limited wireless resources, the server may be only able to serve some particular subsets of clients in each system state. To be more specific, we denote the system state in the kth time interval by $c(k) \in C$, where C is a finite set, and $\{c(1), c(2), \ldots\}$ are i.i.d. random variables with $Prob\{c(k) = c\} =: p_c$. In practice, not only the system state but also the distribution of system states can be time-varying. However, the distribution of system states usually evolves on a much slower time scale compared to the length of a time interval and thus is assumed to be static.

A subset S of clients is said to be *schedulable* under system state c if it is possible for the server to serve all clients in S. For simplicity, we represent a system state c by the collection of subsets S that are schedulable under c. Thus, we have $S \in c$ if S is schedulable under c, and $S \notin c$ otherwise. Since the constraints of schedulable sets can be defined arbitrarily, this model can be applied to a wide range of applications. We will illustrate some examples of applications in Section 6.2. In particular, it can accommodate per-packet delay constraints and rate adaptation.

The server is in charge of choosing a schedulable subset $S \in c(k)$ to serve in each time interval k. The server's choice is described by a *scheduling policy*.

Definition 6.1   *Let $h(k)$ be the system's history up to the kth time interval. A scheduling policy is a function $\eta : (h(k-1), c(k)) \mapsto 2^{\{1,2,...,N\}}$, such that given history $h(k-1)$ and current system state $c(k)$, the server chooses the subset $\eta[h(k-1), c(k)] \in c(k)$ of clients to serve. All clients $n \in \eta[h(k-1), c(k)]$ are considered to be* served in the kth time interval.

We measure the performance of a client through its average rate of being served. We define the *service rate* of a client n as follows:

Definition 6.2   *Let $q_n(k)$ denote the* service rate *of client n up to the kth time interval, defined by the recursion:*

$$q_n(k+1) = \begin{cases} (1 - \frac{1}{k})q_n(k) + \frac{1}{k}, & \text{if client } n \text{ is served at the } k^{th} \text{ interval,} \\ \\ (1 - \frac{1}{k})q_n(k), & \text{otherwise.} \end{cases}$$

The long-term service rate *of client n is defined as $q_n := \liminf_{k \to \infty} q_n(k)$.*

We further assume that each client n has a *utility function* $U_n(\cdot)$. Similar to those discussed in Chapter 5, the utility functions are strictly increasing, strictly concave, and infinitely differentiable. Suppose the system terminates at the end of the kth time interval, client n receives utility that is equivalent to an amount $kU_n(q_n(k))$ of money, and hence its average utility is $U_n(q_n(k))$ per time interval. The *long-term utility* of client n is defined as $\liminf_{k \to \infty} U_n(q_n(k))$, which equals $U_n(q_n)$ since $U_n(\cdot)$ is continuous.

Finally, to enforce some form of fairness among clients, we also assume that each client n has a requirement of *minimum long-term service rate, $\underline{q}_n$*; that is, it requires $q_n \geq \underline{q}_n$ with probability 1. We assume that the minimum long-term service rate requirements are strictly feasible, that is, there exists some scheduling policy that ensures $q_n > \underline{q}_n$, for all n.

We are interested in maximizing the *total long-term utility* of the network, $\sum_{n=1}^{N} U_n(q_n)$. The utility maximization problem of this framework can hence be expressed as:

$$\text{Max} \sum_{n=1}^{N} U_n(q_n)$$

s.t. Network dynamics and schedulability constraints,
and $q_n \geq \underline{q}_n, \forall n$.

However, this formulation only considers the long-term behavior of the system. A solution to this utility maximization problem may not translate into an implementable scheduling policy, which would have to make decisions on a per-interval basis. Thus, we also wish to design *utility-optimal* scheduling policies.

Definition 6.3   *A scheduling policy η is said to be* utility-optimal *if, by applying η, $\sum_{n=1}^{N} U_n(q_n(k))$ converges to the optimal value of the utility maximization problem almost surely as $k \to \infty$.*

6.2 EXAMPLES OF APPLICATIONS

We will first discuss several applications that can be described by our framework.

6.2.1 DELAY-CONSTRAINED WIRELESS NETWORKS WITH RATE ADAPTATION

We first show that the model introduced in Chapter 4 is a special case of the one used in this chapter when rate adaptation is employed. Assume that there are N wireless clients and one access point (AP). Time is assumed to be slotted and divided into time intervals, each consisting of T consecutive time slots. At the beginning of each time interval, packets for each client arrive at the AP. Each client specifies a delay bound of τ_n time slots, with $\tau_n \leq T$. The packet for client n is to be delivered no later than the τ_n^{th} time slot in each time interval. Otherwise, the packet expires and is dropped from the system.

Due to channel fading, the link qualities between the AP and the client can be time-varying. We assume that the AP has full knowledge of the current channel state. The AP then applies rate adaptation for error-free transmissions. Thus, the transmission rates for different clients can be different, which in turn results in different transmission times. We define $t_{c,n}$ as the number of time slots required for an error-free transmission for client n under system state c. A scheduling policy is one which selects an ordered subset $S = \{s_1, s_2, \ldots, s_m\}$ of clients and transmits packets for clients in S according to the order. The ordered subset is considered schedulable under system state c if packets for all clients in S can be delivered before their respective delay bounds, or, to be more specific, $\sum_{n=1}^{i} t_{c,s_n} \leq \tau_{s_i}$, for all $1 \leq i \leq m$. In this scenario, the service rate of each client reflects its timely-throughput.

6.2.2 MOBILE CELLULAR NETWORKS

Consider a mobile cellular network with a base station and N users. The system may have more than one channel, but each channel can be occupied by at most one user at any given time. We assume that time is slotted, where a *time slot* corresponds to a *time interval* in the system model. The length of a time slot is defined as the time needed for transmitting a packet plus any control overhead. Also, due to mobility, the link qualities between the base station and a user can be time-varying. We consider an ON/OFF model for links. The link between a user and the base station is considered ON if a packet can be transmitted between the two without errors, and considered OFF otherwise. We assume that the base station never transmits packets to users with OFF links. Thus, the system state at any time slot can be described as the set of users with ON links. A subset S of users is considered schedulable under some system state c if for any user $n \in S$, the link between user n and the base station is ON, and the size of S is smaller than or equal to the number of channels. A scheduling policy is one which chooses, based on current system state and past history, a schedulable subset of users and assigns channels to each of them. Finally, the service rate of each user is equal to its throughput.

6.2.3 DYNAMIC SPECTRUM ALLOCATION

Consider a scenario with one primary user and N secondary users. The primary user holds licenses for several channels over a large geographical region. TV broadcasters are typical examples of primary users. The primary user only uses a portion of its licensed channels and is willing to allocate unused channels to secondary users. The secondary users are scattered throughout the region and constrained to much smaller transmission powers compared to the primary user, which makes spatial reuse possible. Still, some secondary users may interfere with each other and thus cannot be allocated the same channel simultaneously. We use a conflict graph $G = (V, E)$ to represent the interference relations between secondary users, where V is the set of secondary users and there is an edge between two users if they interfere with each other.

The primary user allocates unused channels periodically. Since the network activity of the primary user can be time-varying, the number of unused channels can also be time-varying. A scheduling policy is one which chooses disjoint subsets of secondary users for each unused channel, with the constraint that two users that are assigned the same channel cannot share a link in the conflict graph.

6.3 A UTILITY MAXIMIZATION APPROACH

In this section, we propose a general method for solving the utility maximization problem in time-varying wireless networks with minimum service requirements. We first show that the utility maximization problem can be formulated as a convex programming problem. Although the formulation requires explicit knowledge of the distribution of system states, i.e., the values of probability $[p_c]$, we will show the surprising result that there exists an on-line scheduling policy that does not need any information on the distribution of system states, and is, further, also utility-optimal.

6.3.1 CONVEX PROGRAMMING FORMULATION

Define $p_c(k)$ and $f_{c,S}(k)$, for all system states c and subsets $S \in c$, recursively, as follows:

$$p_c(k+1) = \begin{cases} \frac{k-1}{k} p_c(k) + \frac{1}{k}, & \text{if } c(k) = c, \\ \\ \frac{k-1}{k} p_c(k), & \text{otherwise,} \end{cases}$$

and

$$f_{c,S}(k+1) = \begin{cases} \frac{k-1}{k} f_{c,S}(k) + \frac{1}{k}, & \text{if } c(k) = c \text{ and } S \text{ is scheduled at the } k^{th} \text{ interval,} \\ \\ \frac{k-1}{k} f_{c,S}(k), & \text{otherwise.} \end{cases}$$

These two variables can be thought of as the relative frequencies of occurrence of the system state c and the event that subset S is scheduled under system state c, respectively. Also, we have $\sum_{S \in c} f_{c,S}(k) = p_c(k)$ and $\sum_c \sum_{S:S \in c, n \in S} f_{c,S}(k) = q_n(k)$ for all c and k. For ease of discussion, we only consider scheduling policies where $f_{c,S} := \lim_{k \to \infty} f_{c,S}(k)$ exists for all system states c and subsets S. Thus, we have $\sum_{S \in c} f_{c,S} = p_c$ and $\sum_c \sum_{S:S \in c, n \in S} f_{c,S} = q_n$. The utility maximization problem can be described as the following convex programming problem:

$$\text{Max} \sum_{n=1}^{N} U_n(q_n) = \sum_{n=1}^{N} U_n(\sum_c \sum_{S:S \in c, n \in S} f_{c,S})$$

$$\text{s.t.} \sum_{S \in c} f_{c,S} = p_c, \forall c,$$

$$q_n = \sum_c \sum_{S:S \in c, n \in S} f_{c,S} \geq \underline{q}_n, \forall n,$$

over $f_{c,S} \geq 0$.

While typical techniques for solving a convex programming problem can be applied to solve this utility maximization problem, such solutions may not be directly translatable into a scheduling policy for our time-varying network. Also, a solution based on solving the convex programming problem would require the knowledge of the probability distribution of system states. In practice, this knowledge may not always be available to the server. Thus, a scheduling policy that makes decisions based only on past history and current system state is needed.

6.3.2   AN ON-LINE SCHEDULING POLICY

We now describe an on-line scheduling policy, and prove that it is utility-optimal. This scheduling policy only requires information on the past history and the current system state, and, surprisingly, does not need any knowledge of the actual probability distribution of system states. Thus, it is easily implementable. The scheduling policy is based on dual decomposition, which is similar to the approach used in Lin and Shroff [69], although they do not consider network dynamics.

We assign a Lagrange multiplier λ_n for each constraint $\sum_c \sum_{S:S \in c, n \in S} f_{c,S} \geq \underline{q}_n$. The resulting Lagrangian of the resulting convex programming problem is:

$$L(f, \lambda) = \sum_{n=1}^{N} U_n(\sum_{c,S:S \in c, n \in S} f_{c,S}) + \sum_{n=1}^{N} \lambda_n(\sum_{c,S:S \in c, n \in S} f_{c,S} - \underline{q}_n),$$

where f denotes the vector consisting of $[f_{c,S}]$, for all c and S, and λ denotes the vector $[\lambda_n]$. The dual objective function is:

$$D(\lambda) = \max_{f:f_{c,S} \geq 0; \sum_{S \in c} f_{c,S} = p_c, \forall c} L(f, \lambda).$$

Since the minimum long-term service rate requirements, $[\underline{q}_n]$, are strictly feasible, there exist $[f_{c,S}]$ such that

$$\sum_{S \in c} f_{c,S} = p_c, \text{ and } \sum_c \sum_{S:S \in c, n \in S} f_{c,S} > \underline{q}_n,$$

for all n. By Slater's condition, $\min_\lambda D(\lambda)$ equals the maximum total utility.

Let $\lambda(k) = [\lambda_n(k)]$ denote Lagrange multipliers that are used in the kth period. The maximum total utility can be achieved by solving two subproblems: maximizing

$$\lim_{k \to \infty} E[L(f(k), \lambda)],$$

for any given λ, and minimizing

$$\lim_{k \to \infty} E[D(\lambda(k))].$$

We will refer to these two subproblems as the *primal problem* and *dual problem*, respectively.

We first discuss how to solve the primal problem. Due to the constraint $\sum_{S \in c} f_{c,S} = p_c$, $[f_{c,S}]$ is an optimal solution if and only if $\frac{\partial L}{\partial f_{c,S}} := \sum_{n \in S}(U'_n(q_n) + \lambda_n) = \max_{S' \in c} \frac{\partial L}{\partial f_{c,S'}}$ for every c and S such that $f_{c,S} > 0$. Recall that $U_n(\cdot)$ is strictly concave, and $U'_n(\cdot)$ is a strictly decreasing function. Suppose, at some time interval k with $c(k) = c$, there exists a subset S schedulable under c such that $\sum_{n \in S}(U'_n(q_n(k)) + \lambda_n) > \sum_{n \in S'}(U'_n(q_n(k)) + \lambda_n)$ for all other subsets S' schedulable under c. We wish to narrow the difference between S and all other S'. One obvious choice would be to schedule the subset S in the time interval, so as to increase $q_n(k+1)$ for all $n \in S$, and thus decrease $\sum_{n \in S}(U'_n(q_n(k)) + \lambda_n)$. In fact, as we shall see in the lemma below, selecting the schedulable subset S that maximizes $\sum_{n \in S}(U'_n(q_n(k)) + \lambda_n)$ also points in the steepest ascent direction of L.

Definition 6.4 *Given λ and $f(k)$, a* max-weight scheduling policy *is one that schedules a schedulable subset $S \in c(k)$ that maximizes $\sum_{n \in S}(U'_n(q_n(k)) + \lambda_n)$ in each time interval k.*

Lemma 6.5 *Let $\Delta f(k)$ be the vector consisting of the elements $\Delta f_{c,S}(k) := f_{c,S}(k+1) - f_{c,S}(k)$ for all c and S. Given λ and $f(k)$, the max-weight scheduling policy also maximizes $E[\nabla L(f, \lambda) \cdot \Delta f(k) | f_{c,S}(k)]$.*

Proof. Recall that we have:

$$f_{c,S}(k+1) = \begin{cases} \frac{k-1}{k} f_{c,S}(k) + \frac{1}{k}, & \text{if } c(k) = c \text{ and } S \text{ is scheduled at the } k^{th} \text{ interval,} \\ \\ \frac{k-1}{k} f_{c,S}(k), & \text{otherwise.} \end{cases}$$

Thus, $\Delta f_{c,S}(k) = \frac{1}{k}(1 - f_{c,S}(k))$ if $c(k) = c$ and S is scheduled, and $\Delta f_{c,S}(k) = -\frac{1}{k} f_{c,S}(k)$, otherwise. Let $\hat{f}_{c,S}(k)$ be the probability that $c(k) = c$ and S is scheduled under the max-weight scheduling policy. We then have:

$$E[\nabla L(f, \lambda) \Delta f(k) | f_{c,S}(k)] = \sum_{c,S} E[\frac{\partial L}{\partial f_{c,S}} \Delta f_{c,S}(k) | f_{c,S}(k)]$$

$$= \frac{1}{k}\{\sum_{c,S}[\hat{f}_{c,S}(k) - f_{c,S}(k)][\sum_{n \in S}(U'_n(q_n(k)) + \lambda_n)]\}. \tag{6.1}$$

Since $Prob\{c(k) = c\} = p_c$, $\sum_S \hat{f}_{c,S}(k) = p_c$. The term $E[\nabla L(f, \lambda)\Delta f(k)|f_{c,S}(k)]$ is maximized by setting:

$$\hat{f}_{c,S}(k) = \begin{cases} p_c, & \text{if } S = \arg\max_{S \in c} \sum_{n \in S} U'_n(q_n(k)) + \lambda_n, \\ 0, & \text{otherwise.} \end{cases} \tag{6.2}$$

This is achieved by selecting the schedulable subset S that maximizes $\sum_{n \in S}(U'_n(q_n(k)) + \lambda_n)$ for every system state c. □

Next, we prove that the max-weight scheduling policy solves the primal problem.

Theorem 6.6   *Under the max-weight scheduling policy,*

$$L(f(k), \lambda) \to D(\lambda), \text{ as } k \to \infty,$$

for any given λ.

Proof. Since the utility functions are infinitely differentiable, $L(f, \lambda)$ is also infinitely differentiable. By Taylor's theorem, we have that for any f, Δf, and fixed λ,

$$L(f + \Delta f, \lambda) = L(f, \lambda) + \nabla L(f, \lambda)\Delta f + r(f, \Delta f, \lambda),$$

where $|r(f, \Delta f, \lambda)| < a(\lambda)|\Delta f|^2$, for some constant $a(\lambda)$. Now we have,

$$\begin{aligned} &E[L(f(k+1), \lambda)|f(k)] \\ \geq\ & L(f(k), \lambda) + E[\nabla L(f(k), \lambda)\Delta f(k) - a(\lambda)|\Delta f(k)|^2 | f(k)] \\ \geq\ & L(f(k), \lambda) + E[\nabla L(f(k), \lambda)\Delta f(k)|f(k)] - \tilde{a}/k^2, \end{aligned} \tag{6.3}$$

where $\Delta f(k)$ is defined as in Lemma 6.5 and $\tilde{a}$ is some constant. The last inequality follows because $|\Delta f_{c,S}(k)| \leq \frac{1}{k}$ for all c, S.

Let $p_c(k) := \sum_{S \in c} f_{c,S}(k)$, which is the empirical frequency that system state c occurs, and let $\hat{f}_{c,S}(k)$ be defined as in the proof of Lemma 6.5. The values of $\hat{f}_{c,S}(k)$ under the max-weight scheduling policy are given as in (6.2). Further, let $\hat{\mu}_c(k) := \max_{S \in c} \sum_{n \in S}(U'_n(q_n(k)) + \lambda_n)$, for all c. Using (6.1) and (6.2),

$$E[\nabla L(f(k), \lambda)\Delta f(k)|f(k)] \geq \frac{1}{k}\sum_c (p_c - p_c(k))\hat{\mu}_c(k).$$

Since $kp_c(k+1)$ is the number of occurrences of system state c until the kth time interval, and the system state in each time interval is i.i.d. distributed, by the law of iterated logarithm [70], there exists some positive constant b such that $\lim\sup_{k\to\infty} \frac{k(p_c(k+1)-p_c)}{k^{1/2}(\log\log k)^{1/2}} \leq b$. Thus, for large enough k, there exists constant $\tilde{b}$ such that

$$E[\nabla L(f(k), \lambda)\Delta f(k)|f(k)] \geq -\frac{(\log\log k)^{1/2}}{k^{3/2}}\tilde{b}.$$

For large enough k, (6.3) can hence be bounded by

$$E[L(f(k+1), \lambda)|f(k)] \geq L(f(k), \lambda) - \frac{(\log \log k)^{1/2}}{k^{3/2}} \tilde{b} - \frac{\tilde{a}}{k^2}. \tag{6.4}$$

As we can see in the above, $E[L(f(k+1), \lambda)|f(k)]$ is "almost" larger than $L(f(k), \lambda)$ except for two diminishing terms. For large enough constant d, $-L(f, \lambda) + d$ is also nonnegative for all f, and by (6.4) it is therefore a "near positive submartingale" as in [71]. Since $\sum_{k=1}^{\infty} [\frac{(\log \log k)^{1/2}}{k^{3/2}} \tilde{b} + \frac{\tilde{a}}{k^2}] < \infty$, Exercise II-4 in [71] shows that $L(f(k), \lambda)$ converges almost surely.

Next, we need to show that $\lim_{k \to \infty} L(f(k), \lambda) = D(\lambda)$. We prove this by contradiction. Recall that the necessary and sufficient condition for $L(f, \lambda) = D(\lambda)$ is that $\sum_{n \in S} (U_n'(q_n) + \lambda_n) = \max_{S' \in c} \sum_{n \in S'} (U_n'(q_n) + \lambda_n)$, for all c, S such that $f_{c,S} > 0$. Suppose $L(f(k), \lambda)$ does not converge to $D(\lambda)$. Then, there exists $\delta > 0$, $\epsilon > 0$ such that for all large enough k, there exist (c_k, S_k) so that $f_{c_k, S_k} > \delta$ and $\sum_{n \in S_k} (U_n'(q_n) + \lambda_n) < \max_{S' \in c} \sum_{n \in S'} (U_n'(q_n) + \lambda_n) - \epsilon$. Evaluating the term $E[\nabla L(f(k), \lambda) \Delta f(k)|f(k)]$ under this condition shows that $E[\nabla L(f(k), \lambda) \Delta f(k)|f(k)] > \frac{1}{k} \delta \epsilon$. Since there exists some constant K such that for all $k > K$, $\frac{\tilde{a}}{k^2} < \frac{1}{k} \delta \epsilon / 2$, we obtain $E[L(f(k+1), \lambda)|f(k)] > L(f(k), \lambda) + \frac{1}{k} \delta \epsilon - \frac{\tilde{a}}{k^2} > L(f(k), \lambda) + \frac{1}{k} \delta \epsilon / 2$. Since $\sum_{k=1}^{\infty} \frac{1}{k} = \infty$, we also have

$$\lim_{k \to \infty} E[L(f(k), \lambda)] = \infty,$$

which is a contradiction. Thus, $\lim_{k \to \infty} L(f(k), \lambda) = D(\lambda)$. □

Next we discuss how to solve the dual problem: $\min_\lambda D(\lambda)$. We use the subgradient method to solve it. We first find a subgradient for $D(\lambda)$.

Lemma 6.7   Let $v_n := [\sum_{c,S:S \in c, n \in S} f_{c,S}^* - \underline{q}_n]$, where $[f_{c,S}^*]$ maximizes $L(f, \lambda)$. Then v is a subgradient of $D(\lambda)$.

Proof. Let λ' be an arbitrary vector. We have:

$$D(\lambda') = \max_{f : \sum_{S \in c} f_{c,S} = p_c, \forall c} L(f, \lambda')$$
$$\geq L(f^*, \lambda') = L(f^*, \lambda) + (\lambda' - \lambda)^T v D(\lambda) = D(\lambda) + (\lambda' - \lambda)^T v.$$

Thus, v is a subgradient of $D(\lambda)$. □

The following theorem then follows from Theorem 8.9.2 in [68]:

Theorem 6.8   Let $\{\beta_k\}$ be a sequence of nonnegative numbers with $\sum_{k=1}^{\infty} \beta_k = \infty$ and $\lim_{k \to \infty} \beta_k = 0$. Update $\lambda(k)$ by:

$$\lambda_n(k+1) = \{\lambda_n(k) - \beta_k [\sum_{c,S:S \in c, n \in S} f_{c,S}^*(k) - \underline{q}_n]\}^+,$$

where $[f_{c,S}^*(k)]$ *maximizes* $L(f, \lambda(k))$. *Then,*

$$\lim_{k \to \infty} D(\lambda(k)) = \min_{\lambda \geq 0} D(\lambda).$$

In practice, the max-weight scheduling policy may converge slowly. Thus, the values of $\lambda(k)$ should be updated at a slower time scale only after the max-weight scheduling policy converges. As $q_n(k) = \sum_{c,S:S \in c, n \in S} f_{c,S}^*(k)$ when the max-weight scheduling policy converges, the subgradient method can be further simplified as

$$\lambda_n(k+1) = \{\lambda_n(k) - \beta_k[q_n(k) - \underline{q}_n]\}^+ \tag{6.5}$$

when updating $\lambda(k)$. In practice, we will update the values of $\lambda(k)$ infrequently. A scheduling policy that jointly applies the max-weight scheduling policy and updates $\lambda(k)$ according to the subgradient method is utility-optimal. This policy is an on-line policy in the sense that it schedules clients and updates $\lambda(k+1)$ only based on $[U_n'(q_n(k))]$, $[q_n(k)]$ and $\lambda(k)$.

6.4   INCENTIVE COMPATIBLE AUCTION DESIGN

We have proposed an on-line scheduling policy and proved that it is utility-optimal in Section 6.3. However, this policy requires knowledge of the utility functions of all clients. In practice, utility functions may be known only to their clients. A strategic client may hence improve its own utility by faking its utility function. In this section, we will propose an auction design that prevents clients from benefiting by faking their utility functions. In this auction, clients offer their bids for service in each time interval. The server selects a subset of clients to serve and charges them based on their bids. The decision of selecting clients and charging them is derived from an auction design similar to the VCG auction [72, 73, 74]. We prove that this design restricts clients from faking their utility functions. We also show that the auction design schedules the same clients as those scheduled by the on-line scheduling policy introduced in Section 6.3. Thus, this auction is not only incentive compatible but also utility-optimal.

6.4.1   BASIC MECHANISM AND INCENTIVE COMPATIBILITY PROPERTY

We first describe the procedure and terminology of an auction. We then propose a VCG-based auction. Appendix B provides a more detailed discussions on the motivation and design of the VCG auction. At the beginning of each time interval k, every client n announces a bid $b_n(k)$ to the server. Based on the bids $[b_n(k)]$, and on the past history of the system and the current system state, the server schedules a subset S in the time interval k and charges each client n an amount $e_n(k)$. Thus, an auction design can be specified by its decisions on selecting clients to schedule and charging them.

Recall that client n receives a utility that is equivalent to an amount $kU_n(q_n(k))$, if the system terminates at the end of the kth time interval. The *net utility* of a client can be defined as

$kU_n(q_n(k)) - e_n(k)$. We can also define the *marginal utility* of client n from being scheduled in the kth time interval as follows:

Definition 6.9 *Let $q_n^+(k)$ and $q_n^-(k)$ be the service rates of client n if it is scheduled, and if it is not scheduled, respectively, in the kth time interval. The* marginal utility *of client n in the kth time interval is defined as $k[U_n(q_n^+(k)) - U_n(q_n^-(k))]$.*

Suppose the goal of every client is to selfishly maximize its own net utility $kU_n(q_n(k)) - e_n(k)$, in each time interval k. An auction design is considered *incentive compatible* if all clients bid their marginal utilities.

Definition 6.10 *An auction design is* incentive compatible *if choosing*

$$b_n(k) = k[U_n(q_n^+(k)) - U_n(q_n^-(k))]$$

yields the highest net utility for client n in time interval k.

Now we propose a VCG-based auction. The server assigns a non-negative *discount*, $d_n(k)$, to each client n in time interval k. The values of discounts are announced before clients offer their bids and are thus not influenced by the bids of clients. After gathering the bids from clients, the server then schedules a schedulable subset S that maximizes $\sum_{n \in S}(b_n(k) + d_n(k))$, with ties broken arbitrarily. The server does not charge anything to those clients that are not scheduled. For each scheduled client $m \in S$, the server charges it the minimum bid $e_m(k)$ it should have offered to be scheduled, specifically

$$e_m(k) = \max_{S':m \notin S'} [\sum_{n \in S'}(b_n(k) + d_n(k))] - \sum_{n \in S, n \neq m}(b_n(k) + d_n(k)) - d_m(k). \qquad (6.6)$$

This auction mechanism is usually referred to as a *weighted VCG* mechanism. The proof that such an auction mechanism is incentive compatible can be found in Section 9.5.3 of [75].

Theorem 6.11 *The proposed auction is incentive compatible.*

Proof. We need to prove that the net utility of client m is maximized when it bids its own marginal utility, given the discounts $[d_n(k)]$. Let $b_m(k) := k[U_m(q_m^+(k)) - U_m(q_m^-(k))]$ be the marginal utility of client m, and let $b'_m(k)$ be any other bid. We show that bidding $b'_m(k)$, instead of $b_m(k)$, cannot increase the net utility for client m.

First consider the case that client m would be scheduled by bidding $b_m(k)$. Let $S, m \in S$, be the subset of clients that would be scheduled when client m bids $b_m(k)$. In this case, we have $\sum_{n \in S}(b_n(k) + d_n(k)) \geq \max_{S':m \notin S'}[\sum_{n \in S'}(b_n(k) + d_n(k))]$. Suppose that client m is also scheduled by bidding $b'_m(k)$. Since the charge for client m for being scheduled, $\max_{S':m \notin S'}[\sum_{n \in S'}(b_n(k) + d_n(k))] - \sum_{n \in S, n \neq m}(b_n(k) + d_n(k)) - d_m(k)$, does not depend on its bid, client m would be charged the same as if it bids $b_m(k)$. Thus, the net utility of client m is not increased. On the other hand, if client is not scheduled by bidding $b'_m(k)$, its net utility will be $kU_n(q_n^-(k))$. However, we have that:

$$\sum_{n \in S}(b_n(k) + d_n(k)) \geq \max_{S':m \notin S'}[\sum_{n \in S'}(b_n(k) + d_n(k))] \tag{6.7}$$

$$\Rightarrow b_m(k) - \{\max_{S':m \notin S'}[\sum_{n \in S'}(b_n(k) + d_n(k))] - \sum_{n \in S, n \neq m}(b_n(k) + d_n(k)) - d_m(k)\} \geq 0 \tag{6.8}$$

$$\Rightarrow kU_n(q_n^+(k)) - e_m(k) \geq kU_n(q_n^-(k)), \tag{6.9}$$

where the left-hand side and the right-hand side of the last inequality are the net utilities of client m if it is scheduled and if it is not, respectively. Thus, the net utility of client m is not increased by bidding $b'_m(k)$.

Next we consider the case when client m would not be scheduled by bidding $b_m(k)$. Thus, there exists a subset S' that does not contain m such that $\sum_{n \in S'}(b_n(k) + d_n(k)) \geq \sum_{n \in S}(b_n(k) + d_n(k))$, for all $S \ni m$. Suppose client m is not scheduled by bidding $b'_m(k)$, then the net utilities for client m under these two bids are identical. On the other hand, suppose a subset $S \ni m$ is scheduled when client m bids $b'_m(k)$; then we have:

$$\max_{S':m \notin S'}[\sum_{n \in S'}(b_n(k) + d_n(k))] \geq \sum_{n \in S}(b_n(k) + d_n(k)) \tag{6.10}$$

$$\Rightarrow b_m(k) - e_m(k) \leq 0 \tag{6.11}$$

$$\Rightarrow kU_n(q_n^-(k)) \geq kU_n(q_n^+(k)) - e_m(k), \tag{6.12}$$

where the left-hand side and right-hand side of the last inequality are the net utilities for client m by bidding $b_m(k)$ and $b'_m(k)$, respectively. Thus, the net utility of client m is not increased by bidding $b'_m(k)$.

We have therefore proved that the proposed auction design is incentive compatible. $\square$

6.4.2 PROOF OF OPTIMALITY

We now prove that the scheduling policy derived from the proposed auction design can be consistent with the max-weight scheduling policy. Thus, this auction also achieves the maximum total long-term utility.

Theorem 6.12 *Let $d_n(k) \equiv \lambda_n(k)$. Assume all clients bid their marginal utility. Then, there exists $\epsilon > 0$ such that if $k > 1/\epsilon$, a schedulable subset $S \in c(k)$ that maximizes $\sum_{n \in S}(b_n(k) + d_n(k))$ also maximizes $\sum_{n \subset S}(U'_n(q_n(k)) + \lambda_n(k))$ over all schedulable subsets.*

Proof. Let $M := \max_{S \in c(k)} \sum_{n \in S}(U'_n(k) + \lambda_n(k))$, and let $\mathbb{S}_M$ be the collection of all schedulable subsets S with

$$\sum_{n \in S}(U'_n(k) + \lambda_n(k)) = M.$$

Also, let $M^- := \max_{S:S \in c(k), S \notin \mathbb{S}_M} \sum_{n \in S} (U_n'(k) + \lambda_n(k))$, and let $\delta := M - M^-$.

Recall that $q_n^+(k) = \frac{k-1}{k} q_n(k-1) + \frac{1}{k}$ and $q_n^-(k) = \frac{k-1}{k} q_n(k-1)$. Since utility functions are infinitely differentiable, by using Taylor's series,

$$b_n(k) = k[U_n(q_n^+(k)) - U_n(q_n^-(k))] = U_n'(\frac{k-1}{k} q_n(k-1)) + O(\frac{1}{k}). \tag{6.13}$$

There exists some ϵ such that if $k > \frac{1}{\epsilon}$,

$$|b_n(k) - U_n'(q_n(k-1))| < \delta/2N. \tag{6.14}$$

Now we have that $| \sum_{n \in S}(b_n(k) + d_n(k)) - \sum_{n \in S}(U_n'(k) + \lambda_n(k))| < \frac{\delta}{2}$ for all subsets S. Thus, for all schedulable subsets $S \in \mathbb{S}_M$ and $S' \notin \mathbb{S}_M$,

$$\sum_{n \in S}(b_n(k) + d_n(k)) > \sum_{n \in S'}(b_n(k) + d_n(k)). \tag{6.15}$$

Therefore, a schedulable subset that maximizes $\sum_{n \in S}(b_n(k) + d_n(k))$ also maximizes $\sum_{n \in S}(U_n'(q_n(k)) + \lambda_n(k))$ □

We have proved that the max-weight scheduling policy and the auction mechanism schedule the same set of clients if $k > \frac{1}{\epsilon}$. Thus, as $k \to \infty$, the two policies will have the same behavior. In Section 6.3.2, we have shown that by applying the max-weight scheduling policy and updating $[\lambda_n(k)]$ according to (6.5), the total long-term utility is maximized. Thus, we can also maximize the total long-term utility by applying the proposed auction and setting discounts $[d_n(k)] \equiv [\lambda_n(k)]$.

6.4.3   IMPLEMENTATION ISSUES

In practice, we implement a protocol that jointly uses the proposed auction design and updates discounts by $[d_n(k)] = [\lambda_n(k)]$. To reduce overhead, all clients announce their utility functions to the server upon joining the system. In each time interval, the server computes their bids as

$$b_n(k) = k[U_n(q_n^+(k)) - U_n(q_n^-(k))]. \tag{6.16}$$

The server then schedules the schedulable subset $S \in c(k)$ that maximizes $\sum_{n \in S}[b_n(k) + d_n(k)]$ and charges all clients $n \in S$ according to (6.6). As a consequence of Theorem 6.11, a client that aims to greedily maximize its own net utility in each time interval k would not benefit by lying about its utility function. Also, this protocol is utility-optimal.

In addition to being incentive compatible and utility-optimal, this protocol also provides incentives for servers to offer service by charging clients. For example, consider the scenario where a TV broadcast company holds several licenses for channels. Even though the company may not fully utilize its channels, it would not be willing to allocate unused bandwidth to unaffiliated users unless doing so can increase its own revenue. Thus, generating revenues for the server is also an important property of a protocol.

6.5   ALGORITHMS FOR SPECIFIC APPLICATIONS

We have shown that the protocol described in Section 6.4.3 is both incentive compatible and utility-optimal. Given the bids $[b_n(k)]$ from clients, the protocol needs to select a subset $S \in c(k)$ that maximizes $\sum_{n \in S} [b_n(k) + d_n(k)]$ and charge them. In this section, we discuss how to design algorithms for scheduling and charging, explicitly for each of the three applications discussed in Section 6.2.

6.5.1   DELAY-CONSTRAINED WIRELESS NETWORKS WITH RATE ADAPTATION

We first consider the scenario described in Section 6.2.1. There are N wireless clients and one AP. Each time interval contains T time slots. At the beginning of each time interval, there is one arrived packet at the AP for each client n. The packet for client n is to be delivered before the τ_n^{th} time slot, with $\tau_n \leq T$, or else the packet expires and is dropped from the system. The transmission time for client n under system state c is $t_{c,n}$.

A schedulable subset $S \in c$ can be described as an ordered subset $S = \{s_1, s_2, \ldots, s_m\}$, such that transmitting packets for clients in S according to the order will meet their respective delay bounds; that is, $\sum_{j=1}^{i} t_{c,s_j} \leq \tau_{s_i}$, for all $1 \leq j \leq m$. To find the schedulable subset $S \in c(k)$ that maximizes $\sum_{n \in S} (b_n(k) + d_n(k))$ in the kth time interval, we can assign a value $v_n(k) := b_n(k) + d_n(k)$ to each client k. Now, if we have $\tau_n \equiv T$, for all n, then this is simply a knapsack problem. To deal with the case when different clients may require different delay bounds, we note that for any schedulable ordered subset S, reordering clients in S in ascending order of their delay bounds, i.e., serving clients in an "earliest deadline first" fashion, is also schedulable. Thus, an analog to the modified knapsack algorithm described in Section 4.4 finds a schedulable subset $S \in c(k)$ that maximizes $\sum_{n \in S} (b_n(k) + d_n(k))$ in the kth time interval. The complete algorithm is shown in Algorithm 4. The complexity of this algorithm is $O(NT)$. To compute the charge for a client $n \in S$, we need to determine $\max_{S' : S' \in c(k), n \notin S'} (b_n(k) + d_n(k))$, which can be obtained by eliminating client n and rerunning Algorithm 4. Thus, the complexity of computing charges for all scheduled clients is $O(N^2 T)$.

6.5.2   MOBILE CELLULAR NETWORKS

Consider the scenario described in Section 6.2.2 with one base station holding $\mathbb{C}$ channels and N mobile users. The links between the users and the base station can either be ON or OFF, and the base station can only schedule transmissions to users with ON links. Recall that a subset S is schedulable under system state c if every client n in S has an ON link, and the size of S is smaller or equal to the number of channels, $\mathbb{C}$. Thus, to implement the protocol, we can simply sort all clients with ON links in descending order of $b_n(k) + d_n(k)$, and schedule the first $\mathbb{C}$ clients. Also, let client m be the $(\mathbb{C} + 1)^{th}$ client with an ON link. To be scheduled, a scheduled client n would have to outbid

---

Algorithm 4 Delay-Constrained Networks

1: **for** $n = 1$ to N **do**
2:     $v_n(k) \leftarrow b_n(k) + d_n(k)$
3: **end for**
4: Sort clients such that $\tau_1 \leq \tau_2 \leq \cdots \leq \tau_N$
5: $S[0, 0] \leftarrow \phi$
6: $M[0, 0] \leftarrow 0$
7: **for** $n = 1$ to N **do**
8:     **for** $t = 1$ to T **do**
9:         **if** $t > \tau_n$ **then**
10:             $M[n, t] \leftarrow M[n, t - 1]$
11:             $S[n, t] \leftarrow S[n, t - 1]$
12:         **else if** $v_n(k) + M[n - 1, t - t_{c(k),n}] > M[n - 1, t]$ **then**
13:             $M[n, t] \leftarrow v_n(k) + M[n - 1, t - t_{c(k),n}]$
14:             $S[n, t] \leftarrow S[n - 1, t - t_{c(k),n}] + \{n\}$
15:         **else**
16:             $M[n, t] \leftarrow M[n - 1, t]$
17:             $S[n, t] = S[n - 1, t]$
18:         **end if**
19:     **end for**
20: **end for**
21: $\max_{S:S \in c(k)} \sum_{n \in S}(b_n(k) + d_n(k)) \leftarrow M(N, T)$
22: schedule according to $S[N, T]$

---

client m, that is, $b_n(k) + d_n(k) \geq b_m(k) + d_m(k)$. Thus, the price paid by each scheduled client n is $b_m(k) + d_m(k) - d_n(k)$.

6.5.3 DYNAMIC SPECTRUM ALLOCATION

Finally, we discuss the scenario of dynamic spectrum allocation. Suppose there is a primary user holding licenses to several channels, and there are N secondary users. The interference relations between secondary users are represented by a conflict graph $G = \{V, E\}$, where V is the set of all secondary users, and there is an edge between two users if they interfere with each other. Suppose the primary user is going to allocate $\mathbb{C}(k)$ unused channels in the kth time interval. A schedulable subset of secondary users can be represented as a coloring by $\mathbb{C}(k)$ colors on some nodes in V, with the restriction that any two colored nodes with the same color cannot share an edge in G. To find a subset $S \in c(k)$ with the maximum $\sum_{n \in S}(b_n(k) + d_n(k))$, we can associate a value $v(k) = b_n(k) + d_n(k)$ to each node in V and find the maximum-weight coloring with $\mathbb{C}(k)$ colors. In particular, when $\mathbb{C}(k) = 1$, this is equivalent to finding the maximum-weight independent set. While finding the maximum-weight independent set is NP-hard, there exist heuristics. Also, for a mid-sized network with about 20 secondary users, the computational overhead for finding an optimal schedule is reasonably small.

CHAPTER 7

Systems with Both Real-Time Flows and Non-Real-Time Flows

In many practical systems, real-time flows and non-real-time flows coexist. In this chapter, we study the problem of maximizing the total utility of non-real-time flows while providing QoS guarantees to real-time flows. We focus on the special case where there is only one AP in the system that has perfect knowledge of channel states. A more generalized setting can be found in [63].

7.1 SYSTEM OVERVIEW AND PROBLEM FORMULATION

Consider a system with one AP and N clients. Time is slotted and time slots are grouped into intervals of length T time slots. Each client is associated with two flows, a real-time flow with inelastic traffic and a non-real-time one with elastic traffic. At the beginning of each interval, the real-time flow of client n generates x_{in} packets, with the requirement that these packets either need to be delivered before the end of the interval, or are otherwise discarded. Each client n has a timely-throughput requirement of q_{in} for its real-time flow. For the non-real-time flows, each client n can determine the number of packets generated by its non-real-time flow in each interval. Packets of non-real-time flows remain in the system until they are delivered. The non-real-time flow of client n is associated with a concave utility function $U_n(\cdot)$. Client n obtains a utility of $U_n(x_{en})$ when the long-term average number of non-real-time packets generated in an interval is x_{en}.

We suppose that the channels are the time-varying. We use c_n to denote the capacity of the link of client n. When n is scheduled in a time slot, it can deliver a total number of c_n packets, including both real-time packets and non-real-time ones. The value of c_n is assumed to be constant in each interval, and is an independent and identically distributed random variable among intervals. We use $c := \{c_1, c_2, \dots\}$ to denote the channel state in an interval.

Let $\mu_{in}(c)$ be the expected number of real-time packets served if the channel state is c, and $\mu_{en}(c)$ be that of non-real-time packets served if the channel state is c. Since the number of real-time packets that can be transmitted in an interval is x_{in}, we require that $\mu_{in}(c) \leq x_{in}$. On the other hand, since non-real-time packets are not dropped at the end of each interval, the number of non-real-time packets for a client can be unbounded. Hence, we do not have such a constraint for $\mu_{en}(c)$.

The long-term average timely-throughput of the real-time flow, and the long-term average throughput of non-real-time flow, of client n can be written as $\mu_{in} := \sum_c \mu_{in}(c)Pr(c)$, and $\mu_{en} := \sum_c \mu_{en}(c)Pr(c)$, respectively. We will focus on the following optimization problem:

$$\text{Max} \sum_n U_n(x_{en}) + w_n\mu_{in} \tag{7.1}$$

$$\text{s.t. } \mu_{in} \geq q_{in}, \forall n, \tag{7.2}$$

$$0 \leq x_{en} \leq \mu_{en}, \forall n, \tag{7.3}$$

$$\mu := \{\mu_{en}, \mu_{in}\} \text{ feasible given channel conditions.} \tag{7.4}$$

$$\tag{7.5}$$

The vector w can be chosen to allocate additional bandwidth to real-time flows beyond their timely-throughput requirements.

7.2   A SOLUTION USING DUAL DECOMPOSITION

Let δ_{in} and δ_{en} be the Lagrangian multipliers for (7.2) and (7.3), respectively. The Lagrangian of (7.1)–(7.4) is

$$L = \sum_n U_n(x_{en}) + w_n\mu_{in} - \delta_{in}(q_{in} - \mu_{in}) - \delta_{en}(x_{en} - \mu_{en}), \tag{7.6}$$

and the dual objective function is

$$D(\delta_e, \delta_i) = \max_{\text{feasible } \mu, \{x_{en}\}} [\sum_n U_n(x_{en}) + w_n\mu_{in} - \delta_{in}(q_{in} - \mu_{in}) - \delta_{en}(x_{en} - \mu_{en})], \tag{7.7}$$

where $\delta_e := \{\delta_{en}\}$ and $\delta_i := \{\delta_{in}\}$. By Slater's condition, we have

$$\min_{\delta_e \geq 0, \delta_i \geq 0} D(\delta_e, \delta_i) = \max \sum_n U_n(x_{en}) + w_n\mu_{in}. \tag{7.8}$$

Moreover, we can rewrite the dual function as

$$D(\delta_e, \delta_i) = \max_{\text{feasible } \mu, \{x_{en}\}} [\sum_n U_n(x_{en}) - \delta_{en}x_{en}]$$
$$+ [\sum_n (w_n + \delta_{in})\mu_{in} + \delta_{en}\mu_{en}]$$
$$- \sum_n \delta_{in}q_{in}. \tag{7.9}$$

Notice that this problem can be decomposed into the following two subproblems:

$$\max_{x_{en}}[U_n(x_{en}) - \delta_{en}x_{en}], \forall n \tag{7.10}$$

and

$$\max_{\text{feasible } \mu} [\sum_n (w_n + \delta_{in})\mu_{in} + \delta_{en}\mu_{en}]. \tag{7.11}$$

Subproblem (7.10) involves finding the optimal x_{en}, which is the rate of the non-real-time flow of client n, to maximize $U_n(x_{en}) - \delta_{en}x_{en}$, and is hence referred to as the *Rate Control Problem*. On the other hand, subproblem (7.11) schedules clients so that the resulting μ maximizes $\sum_n (w_n + \delta_{in})\mu_{in} + \delta_{en}\mu_{en}$, and is referred to as the *Scheduling Problem*.

7.3   A DYNAMIC ALGORITHM AND ITS CONVERGENCE

The decomposition suggests the following iterative algorithm: Let $\delta_{en}(k)$ and $\delta_{in}(k)$ be the Lagrange multipliers used in the k-th interval. In each interval k, each client n generates $x_{en}(k)$ packets to maximize $U_n(x_{en}) - \delta_{en}x_{en}$. That is, each client n chooses

$$x_{en}(k) = \arg\max_{x_{en}} U_n(x_{en}) - \delta_{en}(k)x_{en}. \tag{7.12}$$

On the other hand, in each interval k, given the current channel state c, the AP schedules packets to maximize $\sum_n (w_n + \delta_{in})\mu_{in} + \delta_{en}\mu_{en}$. That is, it chooses a schedule so that the resulting vector of packet deliveries for each flow $\mu(k)$ is

$$\mu(k) = \arg\max_{\mu: \text{ feasible under } c} \sum_n (w_n + \delta_{in}(k))\mu_{in} + \delta_{en}(k)\mu_{en}. \tag{7.13}$$

Since the objective function $\sum_n (w_n + \delta_{in}(k))\mu_{in} + \delta_{en}(k)\mu_{en}$ is a linear equation, $\mu(k)$ can be found using a simple greedy policy.

It is trivial to show that, when $\delta_{en}(k) \equiv \delta_{en}$,

$$\liminf_{K \to \infty} U_n(\frac{\sum_{k=1}^{K} x_{en}(k)}{K}) - \delta_{en}\frac{\sum_{k=1}^{K} x_{en}(k)}{K} = \max_{x_{en}} U_n(x_{en}) - \delta_{en}(k)x_{en}. \tag{7.14}$$

Also, when $\delta_{in}(k) \equiv \delta_{in}$, since $\mu(k)$ maximizes $\sum_n (w_n + \delta_{in})\mu_{in} + \delta_{en}\mu_{en}$ for each given c, we also have

$$E[\sum_n (w_n + \delta_{in})\mu_{in}(k) + \delta_{en}\mu_{en}(k)] = \max_{\text{feasible } \mu} [\sum_n (w_n + \delta_{in})\mu_{in} + \delta_{en}\mu_{en}]. \tag{7.15}$$

In sum, using (7.12) and (7.13) achieves $D(\delta_e, \delta_i)$.

Next, we discuss how to find the optimal Lagrange multipliers to minimize $D(\delta_e, \delta_i)$. They are updated in each interval according to the following:

$$\delta_{in}(k+1) = \{\delta_{in}(k) + \epsilon(q_n - \mu_{in}(k))\}^+, \tag{7.16}$$
$$\delta_{en}(k+1) = \{\delta_{en}(k) + \epsilon(x_{en}(k) - \mu_{en}(k))\}^+. \tag{7.17}$$

As in Theorem 6.8, it can be shown that $D(\delta_e(k), \delta_i(k))$ converges to $\min D(\delta_e, \delta_i)$.

It can be noticed that, if we set $\delta_{in}(0) = \delta_{en}(0) = 0$ for all n, then $\delta_{in}(k)/\epsilon$ is actually the delivery debt, $r_n^{(3)}(k)$, as defined in Example 4.9. Also, $\delta_{en}(k)/\epsilon$ is the queue length of the non-real-time flow of n. Let $l_n(k)$ be the queue length of the non-real-time flow of n in interval k. We can actually rewrite the solutions to the Rate Control Problem and the Scheduling Problem as

$$x_{en}(k) = \underset{0 \leq x_{en} \leq X_{max}}{\arg\max} \frac{1}{\epsilon} U_n(x_{en}) - l_n(k)x_{en}, \tag{7.18}$$

and

$$\mu(k) = \underset{\mu:\ \text{feasible under } c}{\arg\max} \sum_n (\frac{w_n}{\epsilon} + r_n^{(3)}(k))\mu_{in} + l_n(k)\mu_{en}. \tag{7.19}$$

CHAPTER 8

Broadcasting and Network Coding

In this chapter, we consider the problem of broadcasting real-time flows over unreliable wireless links. By broadcasting is that each transmission is intended for multiple receivers. We extend the model introduced in Chapter 2, which only considers unicast flows, to address additional challenges introduces by broadcasting. This model also considers the optional usage of various network coding mechanisms. Based on this model, we propose a framework for designing scheduling policies under different coding mechanisms. We derive scheduling policies for three different systems, one without network coding, one employing XOR coding, and the last employing linear coding.

8.1 SYSTEM MODEL

The model introduced in this section is similar to that in Chapter 2. Two major differences distinguish the two models. First, in the scenario of broadcasting real-time flows, there is usually more than one client that requires packets from the same flow. Second, as ACKs are not implemented for broadcast, and it is costly to obtain feedback information from all clients, an AP that broadcasts several real-time flows does not have per-transmission feedback from clients. For completeness, we formally introduce the model in the sequel.

We consider a wireless system where there is one AP broadcasting several flows with delay constraints to a number of wireless clients. We denote by $\mathbb{I}$ the set of flows, and by $\mathbb{N}$ the set of clients. We assume that time is slotted and numbered as $\tau \in \{0, 1, 2, \dots\}$. The AP can make exactly one transmission in a time slot and the duration of a time slot is hence set to be the time needed for broadcasting one packet. Time slots are divided into *intervals*, where each interval consists of the T consecutive time slots in $[kT, (k + 1)T)$ for $k \geq 0$. According to its specific traffic pattern, each flow may generate at most one packet at the beginning of each interval. We model the traffic patterns of flows as an irreducible Markov chain with finite states and assume that, in the steady state, the probability that the subset $S \subseteq \mathbb{I}$ of flows generates packets is $R(S)$. Each packet generated by any flow has a delay constraint of T time slots; that is, it needs to be delivered to its client within the same interval that it is generated.

We assume that the reliability of the channel between the AP and client n is p_n. Since the overhead of gathering feedbacks from clients is large for broadcast, and ACKs are not implemented for broadcast in most mechanisms, we assume that the AP has no knowledge of whether clients receive the packet correctly after each transmission. This lack of feedback information is one of the most important characteristics that distinguishes this model from that used in Chapter 2 for unicast flows. While it is infeasible for the AP to gather feedbacks from clients on a per-transmission basis, the AP can still obtain feedback infrequently. Such infrequent feedback is used to estimate the channel condition for each client, rather than to acknowledge the reception of a packet. Thus, we assume that the AP has knowledge of the channel reliabilities p_n for all $n \in \mathbb{N}$. Since wireless links are unreliable, the AP may need to broadcast the same packet more than once in an interval to increase the probability of delivery, and thus it is possible that a client receives duplicate packets. In such a case, the duplicate packets are dropped by the client.

Similar to Chapter 2, the performance of client n on flow i is measured by the long-term average number of packets of flow i received by client n per interval, that is, the *timely-throughput* of client n on flow i. Further, we assume that each client n has a specified timely-throughput requirement $q_{i,n}$ for each flow i. The system is considered *fulfilled* if the long-term average number of packets from every flow $i \in \mathbb{I}$ received by client n, excluding duplicate packets, per interval, is at least $q_{i,n}$, for all $n \in \mathbb{N}$.

The goal of this chapter is to design *feasibility-optimal* broadcast policies which fulfill all *strictly feasible* systems. Since the set of broadcast policies depends on the coding mechanism used by the system, we also need to define the concept of *schedule space*.

Definition 8.1   A *schedule space* of a coding mechanism is the collection of all schedules for an interval. For each $S \subseteq I$, it consists of the decision of what packet to transmit in each of the time slots within the interval, which can be carried out by the coding mechanism, given that only flows in S generate a packet at the beginning of the interval.

Definition 8.2   A *broadcast policy* is one that, based on past system history and packet generations in the current interval, assigns a schedule, possibly at random, from the schedule space of its employed coding mechanism.

In this paper, we consider three schedule spaces, one that only transmits raw packets without coding, one that employs XOR coding, and one that employs linear coding.

Definition 8.3   A system is strictly feasible for a schedule space if there exists a positive number $\epsilon > 0$, and a broadcast policy under the schedule space of its coding mechanism, which fulfills the same system with timely-throughput requirement $[(1 + \epsilon)q_{i,n}]$.

Definition 8.4   A broadcast policy is a *feasibility-optimal* policy under the schedule space of some coding mechanism if it fulfills all systems that are strictly feasible under the use of that coding mechanism.

8.2   A FRAMEWORK FOR DESIGNING FEASIBILITY-OPTIMAL POLICIES

We now introduce a framework for designing feasibility-optimal policies under any schedule space. Since the AP does not have feedback information from clients, it cannot know the actual timely-throughput received by each client for each flow. However, it can estimate it. Let $q_{i,n}^*(k)$ be the indicator function that client n actually receives the packet from flow i in the interval $[kT, (k+1)T)$ under some policy η. Note that $q_{i,n}^*(k)$ is a random variable whose value is not known to the server. Let $\hat{q}_{i,n}(k) := \mathrm{E}[q_{i,n}^*(k)|\mathcal{H}_{kT}]$, where $\mathcal{H}_{kT}$ is the history of all packet arrivals of all the flows up to and including time kT, with the conditional expectation taken under the broadcast policy used by the AP. Since the probability of successful reception of a packet by a client depends only on the number of times that a packet is broadcast and is conditionally independent of everything else, it follows that $\hat{q}_{i,n}(k)$ is the conditional probability estimate made by the AP, of whether a packet of flow i is successfully delivered to client n in that interval, based on its actions in that interval. We will denote this aforesaid set of actions of the AP by $\mathcal{A}_{kT}$. So $\hat{q}_{i,n}(k) := \mathrm{E}[q_{i,n}^*(k)|\mathcal{H}_{kT}] = \mathrm{E}[q_{i,n}^*(k)|\mathcal{A}_{kT}]$. The random variables $(\hat{q}_{i,n}(k) - q_{i,n}^*(k))$ are bounded and $E[\hat{q}_{i,n}(k) - q_{i,n}^*(k)|\mathcal{H}_{kT}] = 0$, for all i, n, and k. Define $q_{i,n}^* := \liminf_{K\to\infty} \frac{\sum_{k=0}^{K-1} q_{i,n}^*(k)}{K}$ and $\hat{q}_{i,n} := \liminf_{K\to\infty} \frac{\sum_{k=0}^{K-1} \hat{q}_{i,n}(k)}{K}$. The former is the actual long-term timely-throughput of client n on flow i, while the latter is the asymptotic estimate made by the AP. We then have $\hat{q}_{i,n} = q_{i,n}^*$ almost surely, by the law of large numbers for martingales (Theorem 2.8), and thus a system is fulfilled if and only if $\hat{q}_{i,n} \geq q_{i,n}$ for all i and n.

We define the *expected delivery debt* for each client n and flow i as $d_{i,n}(k) := \sum_{j=0}^{k-1}(q_{i,n} - \hat{q}_{i,n}(j))$, that is, the difference between the number of packets of flow i that should have been delivered to client n to fulfill its timely-throughput requirement, and the expected number of packets of flow i that are delivered to client n, up to time kT, as estimated by the AP. Denote by $D(k)$ the vector consisting of all the expected delivery debts $[d_{i,n}(k)]$. We then have

Lemma 8.5   *A system is fulfilled by a policy η if, under η, $\limsup_{k\to\infty}(\frac{d_{i,n}(k)}{k})^+ = 0$, for all $i \in \mathbb{I}$ and $n \in \mathbb{N}$, where $x^+ := \max\{x, 0\}$.*

We now provide a sufficient condition for a policy to be feasibility-optimal, similar to that used in the proof of Theorem 4.11, and it is also based on the following Theorem 4.10.

Theorem 8.6   *Let S_k be the set of flows that generate packets in the interval $[kT, (k+1)T)$. A scheduling policy η^0 that maximizes*

$$\sum_{i\in\mathbb{I}, n\in\mathbb{N}} d_{i,n}(k)^+ \hat{q}_{i,n}(k) \tag{8.1}$$

for all k, among all policies under its schedule space, is feasibility-optimal for its schedule space.

Proof. Consider a strictly feasible system with timely-throughput requirements $[q_{i,n}]$. There exists a positive number ϵ and a stationary randomized scheduling policy η', which chooses a schedule randomly from the schedule space, based on the packet arrivals at the beginning of this interval and independent of the system history before this interval, that fulfills the same system with timely-throughput requirements $[(1 + \epsilon)q_{i,n}]$. Since we model packet generations in each interval as an irreducible finite-state Markov chain and η' is a stationary randomized policy, there exists a large enough positive number M such that the expected average timely-throughputs under η' in any M consecutive intervals is larger than $(1 + \frac{\epsilon}{2})q_{i,n}$, i.e.,

$$E[\frac{\sum_{l=k}^{k+M-1} \hat{q}_{i,n}(l)}{M}|S_k, D(k)] > (1 + \frac{\epsilon}{2})q_{i,n}, \tag{8.2}$$

for all i, n, S_k, and $D(k)$.

Define $L(t) := \frac{1}{2} \sum_{i\in\mathbb{I},n\in\mathbb{N}}(d_{i,n}(tM)^+)^2$. We then have

$$L(t + 1) = \frac{1}{2} \sum_{i\in\mathbb{I},n\in\mathbb{N}} [(d_{i,n}(tM) + Mq_{i,n} - \sum_{k=tM}^{(t+1)M-1} \hat{q}_{i,n}(k))^+]^2, \tag{8.3}$$

and

$$E[L(t + 1) - L(t)|\mathcal{H}_{tM}] = E[L(t + 1) - L(t)|S_{tM}, D(tM)]$$

$$\leq E[\sum_{i\in\mathbb{I},n\in\mathbb{N}} (Mq_{i,n} - \sum_{k=tM}^{(t+1)M-1} \hat{q}_{i,n}(k))d_{i,n}(tM)^+ + B_0|S_{tM}, D(tM)] \tag{8.4}$$

$$= E\{\sum_{k=tM}^{(t+1)M-1} E[\sum_{i,n} d_{i,n}(k)^+(q_{i,n} - \hat{q}_{i,n}(k))|S_k, D(k)] + B_1(\eta)|S_{tM}, D(tM)\}, \tag{8.5}$$

where B_0 is a positive constant and $B_1(\eta)$ is bounded by $|B_1(\eta)| < B_2$, for some $B_2 > 0$, regardless of S_{tM}, $D(tM)$, and the employed policy η, because $|d_{i,n}(k) - d_{i,n}(tM)| \leq M$, for all $k \in [tM, (t + 1)M)$, and all $i \in \mathbb{I}, n \in \mathbb{N}$.

Let $\hat{q}_{i,n}^0(k)$ and $\hat{q}_{i,n}'(k)$ be the values of $\hat{q}_{i,n}(k)$ under the policies η^0 and η', respectively. Since η^0 maximizes

$$\sum_{i\in\mathbb{I},n\in\mathbb{N}} d_{i,n}(k)^+\hat{q}_{i,n}(k), \tag{8.6}$$

we have that, under η^0,

$$E[L(t+1) - L(t)|S_{tM}, D(tM)] \tag{8.7}$$

$$\leq E\{ \sum_{k=tM}^{(t+1)M-1} E[\sum_{i,n} d_{i,n}(k)^+(q_{i,n} - \hat{q}_{i,n}^0(k))|S_k, D(k)] + B_1(\eta^0)|S_{tM}, D(tM)\} \tag{8.8}$$

$$\leq E\{ \sum_{k=tM}^{(t+1)M-1} E[\sum_{i,n} d_{i,n}(k)^+(q_{i,n} - \hat{q}_{i,n}'(k))|S_k, D(k)] + B_1(\eta^0)|S_{tM}, D(tM)\} \tag{8.9}$$

$$\leq E[M \sum_{i\in\mathbb{I}, n\in\mathbb{N}} (q_{i,n} - \frac{\sum_{k=tM}^{(t+1)M-1} \hat{q}_{i,n}'(k)}{M})d_{i,n}(tM)^+ + B_1(\eta^0) + B_0 - B_1(\eta')|S_{tM}, D(tM)] \tag{8.10}$$

$$< -\frac{M\epsilon q^*}{2} \sum_{i\in\mathbb{I}, n\in\mathbb{N}} d_{i,n}(tM)^+ + B, \tag{8.11}$$

where $q^* := \min_{i,n:q_{i,n}>0} q_{i,n}$ and $B := 2B_2 + B_0$. The last inequality follows from (8.2).

By Theorem 4.10, we have that

$$\limsup_{K\to\infty} \frac{1}{K} \sum_{t=0}^{K-1} E\{ \sum_{i\in\mathbb{I}, n\in\mathbb{N}} d_{i,n}(tM)^+\} \leq \frac{2B}{M\epsilon q^*}. \tag{8.12}$$

Lemma 4.12 shows that $\limsup_{k\to\infty}(\frac{d_{i,n}(k)}{k})^+ = 0$ and the system is also fulfilled by the policy η^0. Thus, η^0 is feasibility-optimal.   □

Theorem 8.6 provides an avenue for designing scheduling policies. For any system and any coding mechanism, we can design a policy that aims to maximize (8.1). Such a policy is feasibility-optimal.

8.3   SCHEDULING WITHOUT NETWORK CODING

Next, we consider three different kinds of coding mechanisms and show how Theorem 8.6 suggests tractable scheduling policies. We first consider a system where network coding is not employed. In each time slot, the AP can only broadcast a raw packet from a flow that has generated one packet in the interval.

Suppose some subset of flows S_k have generated packets at the beginning of the interval $[kT, (k+1)T)$ and that the packet from flow $i, i \in S_k$, is transmitted t_i times within the interval. The probability that client n receives the packet from flow i in this interval is then $\hat{q}_{i,n}(k) = 1 - (1 - p_n)^{t_i}$. Since the AP can make T broadcasts in an interval, we have $\sum_{i\in S_k} t_i \leq T$. We can then formulate the condition in Theorem 8.6 as an integer programming problem:

$$\max \sum_{i \in S_k} \sum_n d_{i,n}(k)^+ [1 - (1 - p_n)^{t_i}] \tag{8.13}$$

$$\text{s.t.} \sum_{i \in S_k} t_i \leq T, \tag{8.14}$$

$$t_i \geq 0, \forall i \in S_k. \tag{8.15}$$

We show that there exists a polynomial time algorithm that solves the integer programming problem. Suppose that, at some time in an interval, the packet of flow i has been broadcast $t_i - 1$ times. The probability that client n has not received the packet from flow i during the first $t_i - 1$ transmissions, and receives this packet when the AP broadcasts the packet from flow i for the t_i-th time, is $p_n(1 - p_n)^{t_i-1}$. Thus, we can define the *weighted marginal delivery probability* of the t_i-th transmission of flow i as

$$m_i(t_i) := \sum_{n \in \mathbb{N}} d_{i,n}(k)^+ p_n(1 - p_n)^{t_i-1}. \tag{8.16}$$

We now propose an online scheduling algorithm, which we call the Greedy Algorithm, as shown in Algorithm 5. In Step 7 of the algorithm, we break ties randomly. We also show that this algorithm is feasibility-optimal.

Theorem 8.7 *Algorithm 5 is feasibility-optimal when network coding is not employed.*

Proof. Suppose that in some interval $[kT, (k + 1)T)$, Algorithm 5 schedules the packet from flow i for transmission t_i^0 times. Suppose there is another algorithm that schedules the packet from flow i for transmission t_i' times, with $\sum_{i \in S_k} t_i' \leq T$. We show that

$$\sum_{i \in S_k, n} d_{i,n}(k)^+ [1 - (1 - p_n)^{t_i^0}] \tag{8.17}$$

$$= \sum_{i \in S_k, n} \sum_{t=1}^{t_i^0} m_i(t) \tag{8.18}$$

$$\geq \sum_{i \in S_k, n} \sum_{t=1}^{t_i'} m_i(t) \tag{8.19}$$

$$= \sum_{i \in S_k, n} d_{i,n}(k)^+ [1 - (1 - p_n)^{t_i'}]. \tag{8.20}$$

If $t_i' \leq t_i^0$, for all $i \in S_k$, then the inequality

$$\sum_{i \in S_k, n} \sum_{t=1}^{t_i^0} m_i(t) \geq \sum_{i \in S_k, n} \sum_{t=1}^{t_i'} m_i(t) \tag{8.21}$$

Algorithm 5 Greedy Algorithm

1: Number flows as $1, 2, \ldots, |\mathbb{I}|$
2: **for** $i = 1$ to $|\mathbb{I}|$ **do**
3:    $t_i \leftarrow 1$
4:    $m_i \leftarrow \sum_{n \in \mathbb{N}} d_{i,n}(k)^+ p_n$
5: **end for**
6: **for** $\tau = 1$ to T **do**
7:    $i \leftarrow \arg\max_{j \in S_k} m_j$
8:    $t_i \leftarrow t_i + 1$
9:    $m_i \leftarrow \sum_{n \in \mathbb{N}} d_{i,n}(k)^+ p_n (1 - p_n)^{t_i - 1}$
10: **end for**
11: **for** $i = 1$ to $|\mathbb{I}|$ **do**
12:    **for** $\tau = 1$ to t_i **do**
13:       broadcast the packet from flow i
14:    **end for**
15:    **for** $n \in \mathbb{N}$ **do**
16:       $d_{i,n}(k+1) \leftarrow d_{i,n}(k) + q_{i,n} - [1 - (1 - p_{b_i,n})^{t_i}]$
17:    **end for**
18: **end for**

---

holds. If there exists some i such that $t'_i > t^0_i$, then there also exists some j such that $t'_j < t^0_j$, since $\sum_{i \subset S_k} t'_i \leq T = \sum_{i \in S_k} t^0_i$. By the design of the algorithm and the fact that both $m_i(\cdot)$ and $m_j(\cdot)$ are decreasing functions, we have that $m_j(t'_i) \leq m_i(t^0_i + 1) \leq m_j(t^0_j) \leq m_j(t'_j + 1)$. Thus, we can decrement t'_i by 1 and increment t'_j by 1 without decreasing the value of $\sum_{i \in S_k, n} \sum_{t=1}^{t'_i} m_i(t)$. We repeat this procedure until $t'_i \leq t^0_i$, for all i, and deduce that $\sum_{i \in S_k, n} \sum_{t=1}^{t^0_i} m_i(t) \geq \sum_{i \in S_k, n} \sum_{t=1}^{t'_i} m_i(t)$. $\square$

We now analyze the complexity of the Greedy Algorithm. We can implement the Greedy Algorithm using a max-heap, where there is one node for each flow i whose value is m_i. In Steps 4 and 8, it takes $O(|\mathbb{N}|)$ time to compute m_i. It takes $O(|\mathbb{I}| \log |\mathbb{I}|)$ time to construct the max-heap. In each iteration between Steps 5 and 8, it takes $O(\log |\mathbb{I}|)$ time to find $\arg\max_{j \in S_k} m_j$ and remove the node from the max-heap. It also takes $O(\log |\mathbb{I}|)$ time to insert that node back into the max-heap once its value is updated. Thus, the total complexity of computing the schedule in an interval is $O(|\mathbb{I}| \log |\mathbb{I}| + T|\mathbb{N}| + T \log |\mathbb{I}|)$. Detailed discussions on max-heap can be found in [76].

8.4   BROADCASTING WITH XOR CODING

In this section, we address the use of XOR coding for broadcasting. We assume that the AP can either broadcast a raw packet from a flow, or it can choose to broadcast an encoded packet (packet from flow i $\oplus$ packet from flow j), the XOR of a packet from flow i with a packet from flow j, which we shall henceforth denote by $i \oplus j$. A client can recover the packet from flow i either upon directly receiving a raw packet from flow i, or upon receiving a raw packet from flow j and an encoded packet $i \oplus j$, for some j. We exhibit a simple example where a system with XOR coding can achieve strictly better performance than one without network coding.

Example 8.8   Consider a system with two flows that generate one packet in each interval, and only one client whose channel reliability is $p_1 = 0.5$. Assume that there are six time slots in an interval. Suppose that the AP transmits each packet three times in an interval. Then we have $\hat{q}_{1,1} = \hat{q}_{2,1} = 0.875$. Thus, a system with timely-throughput requirements $q_{1,1} = q_{2,1} > 0.875$ is not feasible when network coding is not employed. On the other hand, a system that employs XOR coding can transmit each of the three different types of packets, the raw packet from each flow and the encoded packet $1 \oplus 2$, twice in each interval. Now, this client can successfully obtain the packet from flow 1 under two circumstances: First, it receives a transmission containing the raw packet from flow 1. Second, it receives a transmission containing the raw packet from flow 2 and a transmission containing the encoded packet $1 \oplus 2$. With some calculations, we can show that using XOR coding achieves $\hat{q}_{1,1} = \hat{q}_{2,1} = 0.890625$.

While it may be computationally complicated to design a feasibility-optimal scheduling policy when XOR coding is employed, we aim to design a tractable policy that achieves better performance than the Greedy Algorithm in Section 8.3. Suppose the Greedy Algorithm broadcasts the packet from flow i for a total of t_i^G times in an interval. We sort all flows so that $t_1^G \geq t_2^G \geq \ldots$, and enforce the following restrictions on our scheduling policy:

1. In addition to raw packets, we only allow encoded packets of the form $(2i - 1) \oplus (2i)$. The intuition behind this restriction is that we only combine two packets which have each been transmitted a similar number of times under the Greedy Algorithm, which implies that they have similar importance.

2. The total number of transmissions scheduled for the raw packets from flow $2i - 1$ and flow $2i$, as well as the encoded packet $(2i - 1) \oplus (2i)$, equals $t_{2i-1}^G + t_{2i}^G$. The intuition behind this restriction is that we aim to enhance the performance of flows $2i - 1$ and $2i$ by XOR coding without hurting other flows.

We call the above two restrictions the *pairwise combination restriction* and the *transmission conservation restriction for XOR coding*, respectively. Suppose that, under some policy η that follows the above restrictions, the raw packet from flow i is transmitted t_i times, and the encoded packet $(2i - 1) \oplus (2i)$ is transmitted $t_{(2i-1)\oplus(2i)}$ times in the k-th interval. The probability that client n receives the packet from flow $(2i - 1)$ is $\hat{q}_{2i-1,n}(k) = 1 - (1 - p_n)^{t_{2i-1}}[(1 - p_n)^{t_{2i}} + (1 - p_n)^{t_{(2i-1)\oplus(2i)}} - (1 - p_n)^{t_{2i}+t_{(2i-1)\oplus(2i)}}]$. Similarly, we also have $\hat{q}_{2i,n}(k) = 1 - (1 - p_n)^{t_{2i}}[(1 - p_n)^{t_{2i-1}} + (1 - p_n)^{t_{(2i-1)\oplus(2i)}} - (1 - p_n)^{t_{2i-1}+t_{(2i-1)\oplus(2i)}}]$. By Theorem 8.6, designing a feasibility-optimal policy among all policies that follow the above restrictions can be simplified to one of solving the following integer programming problem for all k:

$$\max \sum_{i=1}^{|S_k|/2} \sum_n d_{2i-1,n}(k)^+ \hat{q}_{2i-1,n}(k) + d_{2i,n}(k)^+ \hat{q}_{2i,n}(k)$$

$$\text{s.t. } t_{2i-1} + t_{2i} + t_{(2i-1)\oplus(2i)} = t_{2i-1}^G + t_{2i}^G, \forall 1 \leq i \leq |S_k|/2,$$
$$t_i \geq 0 \qquad\qquad , \forall 1 \leq i \leq |S_k|,$$
$$t_{(2i-1)\oplus(2i)} \geq 0 \qquad\qquad , \forall 1 \leq i \leq |S_k|/2.$$

In the formulation, we assume that $|S_k|$, the number of flows that generate a packet in the k-th interval, is even. If $|S_k|$ is odd, we can add an imaginary flow i^* into the system to make $|S_k|$ even. We set $q_{i^*,n} = 0$, for all n, and thus $t_{i^*}^G = 0$ since it will never be scheduled by the Greedy Algorithm. The condition $t_{2i-1} + t_{2i} + t_{(2i-1)\oplus(2i)} = t_{2i-1}^G + t_{2i}^G$ allows us to decompose this integer programming problem into $|S_k|/2$ subproblems so that the i-th subproblem only involves flows $2i - 1$ and $2i$:

$$\max \sum_n d_{2i-1,n}(k)^+ \hat{q}_{2i-1,n}(k) + d_{2i,n}(k)^+ \hat{q}_{2i,n}(k) \tag{8.22}$$

$$\text{s.t. } t_{2i-1} + t_{2i} + t_{(2i-1)\oplus(2i)} = t_{2i-1}^G + t_{2i}^G, \tag{8.23}$$
$$t_{2i-1}, t_{2i}, t_{(2i-1)\oplus(2i)} \geq 0. \tag{8.24}$$

If we further assume that $t_{(2i-1)\oplus(2i)}$ is fixed, this subproblem is equivalent to

$$\max \sum_n d_{2i-1,n}(k)^+ (1 - p_n)^{t_{(2i-1)\oplus(2i)}}[1 - (1 - p_n)^{t_{2i-1}}]$$

$$+ \sum_n d_{2i,n}(k)^+ (1 - p_n)^{t_{(2i-1)\oplus(2i)}}[1 - (1 - p_n)^{t_{2i}}] + C \tag{8.25}$$

$$\text{s.t. } t_{2i-1} + t_{2i} = t_{2i-1}^G + t_{2i}^G - t_{(2i-1)\oplus(2i)}, \tag{8.26}$$
$$t_{2i-1}, t_{2i} \geq 0, \tag{8.27}$$

where C is a constant. The optimal (t_{2i-1}, t_{2i}) for this problem can be found by Algorithm 6. The complexity of Algorithm 6 is $O(|\mathbb{N}|(t_{2i-1}^G + t_{2i}^G))$. We have the following lemma, whose proof is essentially the same as that of Theorem 8.7.

Lemma 8.9   *Given $t_{(2i-1)\oplus(2i)}$, the pair (t_{2i-1}, t_{2i}) found by Algorithm 6 maximizes*

$$\sum_n d_{2i-1,n}(k)^+ \hat{q}_{2i-1,n}(k) + d_{2i,n}(k)^+ \hat{q}_{2i,n}(k).$$

Algorithm 6 GreedyXOR$(i, t_{(2i-1)\oplus(2i)}, t_{2i-1}^G, t_{2i}^G)$

1: $t_{2i-1} \leftarrow 1$

2: $t_{2i} \leftarrow 1$

3: $m_{2i-1} \leftarrow \sum_{n \in \mathbb{N}} d_{2i-1,n}(k)^+ (1-p_n)^{t_{(2i-1)\oplus(2i)}} p_n$

4: $m_{2i} \leftarrow \sum_{n \in \mathbb{N}} d_{2i,n}(k)^+ (1-p_n)^{t_{(2i-1)\oplus(2i)}} p_n$

5: **for** $\tau = 1$ to $t_{2i-1}^G + t_{2i}^G - t_{(2i-1)\oplus(2i)}$ **do**

6:      $j \leftarrow \arg\max\{m_{2i-1}, m_{2i}\}$

7:      $t_j \leftarrow t_j + 1$

8:      $m_j \leftarrow \sum_{n \in \mathbb{N}} d_{j,n}(k)^+ (1-p_n)^{t_{(2i-1)\oplus(2i)}} p_n (1-p_n)^{t_j-1}$

9: **end for**

10: **return** (t_{2i-1}, t_{2i})

Using Algorithm 6 as a building block, we propose the *Pairwise XOR* algorithm, shown in Algorithm 7, to find the optimal schedule when XOR coding is employed under the two afore-mentioned restrictions. The complexity of the Pairwise XOR algorithm is $O(|\mathbb{I}| \log |\mathbb{I}| + T|\mathbb{N}| + T \log |\mathbb{I}| + \sum_{i=1}^{|S_k|/2} |\mathbb{N}|(t_{2i-1}^G + t_{2i}^G)^2) = O(|\mathbb{I}| \log |\mathbb{I}| + T \log |\mathbb{I}| + T^2|\mathbb{N}|)$. The following theorem is the direct result of Lemma 8.9 and Theorem 8.6.

Theorem 8.10   *The Pairwise XOR algorithm is feasibility-optimal among all policies that follow the pairwise combination restriction and the transmission conservation restriction for XOR coding. In particular, the Pairwise XOR algorithm fulfills every system that can be fulfilled by the Greedy Algorithm.*

8.5   BROADCASTING WITH LINEAR CODING

In this section, we address the use of linear coding to improve the performance of broadcasting delay-constrained flows. We assume that, in addition to raw packets, the AP can also broadcast packets that contain linear combinations of packets from any subset of flows $L \subseteq S_k$. A client can decode all packets from the subset L of flows if it receives at least $|L|$ packets that contain linear combinations of packets from these flows. If a client receives less than $|L|$ packets containing such linear combinations, none of the packets from these flows can be decoded. We first exhibit a simple example where a system that uses linear coding provides better performance than one that does not use network coding.

Example 8.11   Consider a system with one client, whose channel reliability is $p_1 = 0.5$, three flows that generate one packet in each interval, and nine time slots in an interval. A similar argument as that in Example 8.8 shows that $q_{1,1} = q_{2,1} = q_{3,1} > 0.875$ is not feasible when network coding is not employed. On the other hand, if the AP employs linear coding and broadcasts a linear combination

---

Algorithm 7 Pairwise XOR

1: Obtain $t_1^G, t_2^G, \ldots$ from the Greedy Algorithm
2: Sort flows so that $t_1^G \geq t_2^G \geq \ldots$
3: **for** $i = 1$ to $|S_k|/2$ **do**
4:   $Opt \leftarrow -\infty$
5:   **for** $t = 0$ to $t_{2i-1}^G + t_{2i}^G$ **do**
6:     $(t_{2i-1}, t_{2i}) \leftarrow GreedyXOR(i, t_{(2i-1)\oplus(2i)}, t_{2i-1}^G, t_{2i}^G)$
7:     **if** $\sum_n d_{2i-1,n}(k)^+ \hat{q}_{2i-1,n}(k) + d_{2i,n}(k)^+ \hat{q}_{2i,n}(k) > Opt$ **then**
8:       $Opt \leftarrow \sum_n d_{2i-1,n}(k)^+ \hat{q}_{2i-1,n}(k) + d_{2i,n}(k)^+ \hat{q}_{2i,n}(k)$
9:       $t_{2i-1}^X \leftarrow t_{2i-1}$
10:       $t_{2i}^X \leftarrow t_{2i}$
11:       $t_{(2i-1)\oplus(2i)}^X \leftarrow t$
12:     **end if**
13:   **end for**
14: **end for**
15: **for** $i = 1$ to $|S_k|/2$ **do**
16:   **for** $\tau = 1$ to t_{2i-1}^X **do**
17:     Broadcast the packet from flow $2i - 1$
18:   **end for**
19:   **for** $\tau = 1$ to t_{2i}^X **do**
20:     Broadcast the packet from flow $2i$
21:   **end for**
22:   **for** $\tau = 1$ to $t_{(2i-1)\oplus(2i)}^X$ **do**
23:     Broadcast the packet $(2i - 1) \oplus (2i)$
24:   **end for**
25: **end for**

---

of the three flows in each time slot, the client can decode all packets from the three flows if it receives at least three packets out of the nine transmissions in an interval, which has probability 0.91015625.

As in Section 8.4, we address the problem of finding a tractable scheduling policy that achieves better performance than the Greedy Algorithm. Suppose that the Greedy Algorithm schedules t_i^G transmissions for the packet from flow i in some interval. We sort all flows so that $t_1^G \geq t_2^G \geq \ldots$, and enforce the following restrictions:

1. Flows are grouped into subsets as $L_1 = \{1, 2, \ldots, l_1\}$, $L_2 = \{l_1 + 1, \ldots, l_2\}$, $\ldots$. In each time slot, the AP broadcasts a linear combination of packets from flows in one of the subsets $L_1, L_2, \ldots$. The intuition behind this restriction is that we only combine packets that have been scheduled similar numbers of times.

2. The AP broadcasts linear combinations of packets from the subset $L_h = \{l_{h-1} + 1, l_{h-1} + 2, \ldots, l_h\}$ a total number of $\sum_{i=l_{h-1}+1}^{l_h} t_i^G$ times, where we set $l_0 = 0$. The intuition behind this restriction is that we aim to enhance the performance of flows within L_h without hurting other flows.

The two restrictions above are called the *adjacent combination restriction* and the *transmission conservation restriction for linear coding*, respectively.

We define $r_{n,u,v}$ as the probability that client n receives at least u packets successfully out of v transmissions. We can compute $r_{n,u,v}$ for all $n \in \mathbb{N}$, $1 \le u < T$, and $1 \le v \le T$ in $O(|\mathbb{N}|T^2)$ time using the following iteration:

$$
r_{n,u,v} = \begin{cases}
1, & \text{if } u = 0, \\
0, & \text{if } u > 1, v = 0, \\
p_n r_{n,u-1,v-1} + (1 - p_n) r_{n,u,v-1}, & \text{else.}
\end{cases}
$$

If flows are grouped as $L_1, L_2, \ldots$ in the k-th interval, the probability that client n is able to obtain the packet from flow $i \in L_h$ is then $\hat{q}_{i,n}(k) = r_{n,|L_h|,(\sum_{i=l_{h-1}+1}^{l_h} t_i^G)}$. We need to find the optimal way to group flows such that

$$
\sum_{i \in S_k} \sum_n d_{i,n}(k)^+ \hat{q}_{i,n}(k)
$$

is maximized. We solve this problem by dynamic programming. Let $H_{i,j}$ be the optimal way to group flows $i, i + 1, \ldots, j$, if flows within $[i, j]$ are not allowed to be grouped with flows outside $[i, j]$. We represent $H_{i,j}$ by the collection of groups formed by flows within $[i, j]$. We then have that $H_{i,j}$ either contains one single group consisting of all flows within $[i, j]$, or is of the form $H_{i,h} \cup H_{h+1,j}$, for some $i \le h < j$. The optimal way to group all flows, which is $H_{1,|S_k|}$, can be found by dynamic programming as in Algorithm 8. The complexity of Algorithm 8 is $O(|\mathbb{I}| \log |\mathbb{I}| + T|\mathbb{N}| + T \log |\mathbb{I}| + |\mathbb{N}||\mathbb{I}|^3)$. We have the following theorem:

Theorem 8.12 *The Optimal Grouping policy, as described in Algorithm 8, is feasibility-optimal among all policies that follow the adjacent combination restriction and the transmission conservation restriction for linear coding. In particular, the Optimal Grouping policy fulfills all systems that can be fulfilled by the Greedy Algorithm.*

Proof. The first part of the theorem follows from the discussion in the previous paragraph and Theorem 8.6. The second part of the theorem follows because a policy that sets $H_{1,|S_k|} = \{\{1\}, \{2\}, \ldots\}$ in each interval k has the same schedule as that resulting from the Greedy Algorithm. □

Algorithm 8 Optimal Grouping

1: Obtain $t_1^G, t_2^G, \ldots$ from the Greedy Algorithm
2: Sort flows so that $t_1^G \geq t_2^G \geq \ldots$
3: **for** $i = 1$ to $|S_k|$ **do**
4:     $O_{i,i} \leftarrow \sum_n d_{i,n}(k)^+ r_{n,1,t_i^G}$
5:     $H_{i,i} \leftarrow \{\{i\}\}$
6: **end for**
7: **for** $s = 2$ to $|S_k|$ **do**
8:     **for** $i = 1$ to $|S_k| - s + 1$ **do**
9:         $total \leftarrow \sum_{h=i}^{i+s-1} t_h^G$
10:         $O_{i,i+s-1} \leftarrow \sum_{h=i}^{i+s-1} \sum_n d_{h,n}(k)^+ r_{n,s,total}$
11:         $H_{i,i+s-1} \leftarrow \{\{i, i+1, \ldots, i+s-1\}\}$
12:         **for** $j = i$ to $i + s - 2$ **do**
13:             **if** $O_{i,j} + O_{j+1,i+s-1} > O_{i,i+s-1}$ **then**
14:                 $O_{i,i+s-1} = O_{i,j} + O_{j+1,i+s-1}$
15:                 $H_{i,i+s-1} = H_{i,j} \cup H_{j+1,i+s-1}$
16:             **end if**
17:         **end for**
18:     **end for**
19: **end for**
20: Group flows as in $H_{1,|S_k|}$ and broadcast them accordingly

---

8.6   SIMULATION RESULTS

We have implemented the three scheduling algorithms proposed in this paper, namely, the Greedy Algorithm, the Pairwise XOR algorithm, and the Optimal Grouping algorithm, in ns-2. We compare their performances against a round-robin scheduling policy.

We use the Shadowing module in ns-2 to simulate the unreliable wireless links between the AP and clients. In the Shadowing module, the link reliability decreases as the distance between two wireless devices increases. The relation between link reliability and distance is shown in Figure 8.1.

We implement our algorithms based on the IEEE 802.11 standard. Under 802.11, broadcasting a packet with size 160 bytes, which is the size of VoIP packets using the G.711 codec [77], takes about 2 ms. We assume the length of an interval is 40 ms, and hence it consists of 20 time slots.

We consider the scenario where an AP is broadcasting 10 delay-constrained flows to 20 clients that are evenly distributed in a 780×1040 area. We consider two different topologies and timely-throughput requirements of clients. The first one is called the *symmetric topology*. In the symmetric topology, the AP is located at the center of the domain, i.e., at position $(390, 520)$. The timely-throughput requirements of each client are α for flows 1–5, and β for flows 6–10. That is, we set

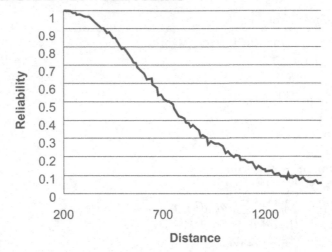

Figure 8.1: Relationship between distance and link reliability.

Table 8.1: timely-throughput requirements of clients in each region for the asymmetric topology

Region	subscribed flows and timely-throughput requirements
$[390, 780] \times [520, 1040]$	$q_{1,n} = q_{5,n} = \alpha, q_{6,n} = q_{10,n} = \beta$
$[390, 780] \times [0, 520]$	$q_{2,n} = q_{5,n} = \alpha, q_{7,n} = q_{10,n} = \beta$
$[0, 390] \times [520, 1040]$	$q_{3,n} = q_{5,n} = \alpha, q_{8,n} = q_{10,n} = \beta$
$[0, 390] \times [0, 520]$	$q_{4,n} = q_{5,n} = \alpha, q_{9,n} = q_{10,n} = \beta$

$q_{i,n} = \alpha$ if $i \leq 5$, and $q_{i,n} = \beta$ if $i > 5$, where α and β are tunable variables to reflect that clients may have different timely-throughput requirements for different flows. The other topology that we consider is called the *asymmetric topology*, where the AP is located at position (520, 650). Further, clients in different regions may subscribe to different flows. The timely-throughput requirements of flows subscribed to by clients in each region are summarized in the above table. We set $q_{i,n} = 0$ if $q_{i,n}$ does not appear in the table.

We study two different types of traffic patterns for packet arrivals, namely, *deterministic arrivals* and *probabilistic arrivals*. For deterministic arrivals, we assume that all the 10 flows generate one packet in each interval. This corresponds to flows carrying constant-bit-rate traffic, such as the G.711 codec for VoIP. For probabilistic arrivals, we assume that each flow generates one packet with some probability, independent of the packet generations of other flows, at the beginning of each interval. This scenario corresponds to flows carrying variable-bit-rate traffic, such as MPEG video streaming. In particular, we assume each of flows 1–5 generates one packet with probability 0.9, and each of flows 6–10 generates one packet with probability 0.6, at the beginning of each interval.

We evaluate the performances of our algorithms and the round-robin policy for each of the two topologies and each of the two traffic patterns. We compare the performances of different scheduling algorithms by comparing the pairs of (α, β) that can be fulfilled by each algorithm. A system is considered to be fulfilled if, after 500 intervals, the average timely-throughput of client n on flow i is at least $q_{i,n} - 0.03$.

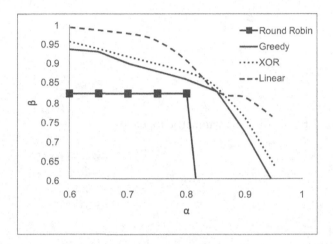

Figure 8.2: Deterministic arrivals and symmetric topology

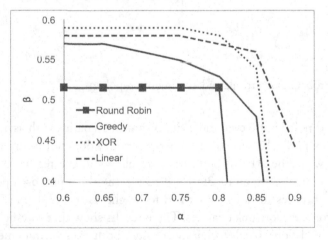

Figure 8.3: Probabilistic arrivals and symmetric topology

The simulation results are shown in Figures 8.2–8.5. As shown in the figures, all the three proposed algorithms outperform the round-robin policy in all scenarios. This is because, without the knowledge of timely-throughput requirements, the round-robin policy cannot offer any tradeoff between flows. Further, the differences of performance between the round-robin policy and the

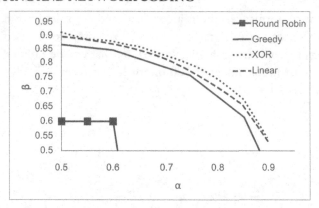

Figure 8.4: Deterministic arrivals and asymmetric topology

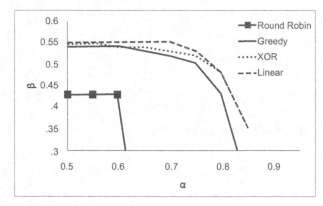

Figure 8.5: Probabilistic arrivals and asymmetric topology

three proposed policies are even more significant in scenarios with asymmetric topology, as shown in Figures 8.4 and 8.5, because the round-robin policy does not incorporate any knowledge of network topology. As the AP has better links to clients in the region $[390, 780] \times [520, 104]$ than those to clients in the region $[0, 390] \times [0, 520]$, the three proposed policies allocate more transmission opportunities to flows subscribed to by clients in the region $[0, 390] \times [0, 520]$ so as to compensate for their low link qualities. These results show that an efficient policy for broadcasting delay-constrained flows needs to jointly consider both client requirements and network topology. Also, both algorithms using network coding achieve better performance than the Greedy Algorithm, which demonstrates that network coding can be used to increase the capacity of wireless networks for broadcasting delay-constrained flows.

We close this section by comparing the Pairwise XOR algorithm for XOR coding and the Optimal Grouping algorithm for linear coding. In the scenario with deterministic arrivals and symmetric topology, the Optimal Grouping algorithm has much better performance than the Pairwise

XOR algorithm. However, the advantage of the Optimal Grouping algorithm becomes less prominent in the other three scenarios, and sometimes it even performs worse than the Pairwise XOR algorithm. The Optimal Grouping algorithm allows combining more than two flows, and thus explores more coding possibilities, which is why it achieves better performance in the first scenario. On the other hand, in systems where the number of generated packets in each interval is low, as in the scenario with probabilistic arrivals, or when the topologies are asymmetric, it becomes less beneficial to combine a large number of packets. In such systems, the Pairwise XOR algorithm may benefit from its simpler coding structure.

APPENDIX A

Lyapunov Analysis and its Application to Queueing Systems

One of the most important tools for studying the stability of scheduling policies is the *Lyapunov analysis*. This tool consists of defining a *Lyapunov function* and studying the *Lyapunov drift*. In this chapter, we introduce this tool and demonstrate its utility by a case study.

Consider a system with N clients that generate some work loads to be served. Let $D_n(t)$ be the service deficiency of client n at time t. In the context of Chapter 4, $D_n(t)$ can be defined as the pseudo-debt of n. We can also consider a queueing system where each client generates packets for a server, and packets that are yet to be served are kept in the queues of their respective clients. At a certain time t, the number of packets in the queue of client n can be written as (Total number of packets generated by n)-(Total number of packets served for n). Thus, in this system, $D_n(t)$ can be defined as the queue length of client n. Our goal is to find a policy that makes the system *strongly stable* whenever possible.

Definition A.1   A system is *strongly stable* if:

$$\limsup_{t \to \infty} \frac{1}{t} \sum_{\tau=0}^{t-1} E\{D_n(\tau)\} < \infty, \tag{A.1}$$

for all n, which also implies that

$$\lim_{t \to \infty} \frac{E\{D_n(t)\}}{t} = 0, \tag{A.2}$$

by Lemma 4.12.

Let $D(t)$ be the vector containing $[D_1(t), D_2(t), \ldots, D_N(t)]$. We define a *quadratic Lyapunov function*, $L(D(t)) := \frac{1}{2} \sum_n D_n(t)^2$, and the *Lyapunov drift* $\Delta(t) := E\{L(D(t+1)) - L(D(t))|D(t)\}$. We then have the following theorem, which is similar to Theorem 4.10.

Theorem A.2   *Suppose there exists some constants $B > 0, \epsilon > 0$ such that:*

$$\Delta(t) \le B - \epsilon \sum_n D_n(t), \tag{A.3}$$

for all t, then:

$$\limsup_{t \to \infty} \frac{1}{t} \sum_{\tau=0}^{t-1} E\{D_n(\tau)\} \leq B/\epsilon, \tag{A.4}$$

and the system is strongly stable.

Proof. Suppose the condition holds. Then

$$E\{L(D(t))\} - E\{L(D(0))\} = E\{\sum_{\tau=0}^{t-1} E\{L(D(\tau+1)) - L(D(\tau))|D(\tau)\}\} \tag{A.5}$$

$$= E\{\sum_{\tau=0}^{t-1} \Delta(\tau)\} \tag{A.6}$$

$$\leq E\{\sum_{\tau=0}^{t-1}(B - \epsilon \sum_{n} D_n(\tau))\} = tB - \epsilon \sum_{\tau=0}^{t-1} E\{D_n(\tau)\}. \tag{A.7}$$

Since $E\{L(D(t))\} \geq 0$, we also have $\frac{1}{t}\sum_{\tau=0}^{t-1} E\{D_n(\tau)\} \leq B/\epsilon + E\{L(D(0))\}/(\epsilon t)$, and hence

$$\limsup_{t \to \infty} \frac{1}{t} \sum_{\tau=0}^{t-1} E\{D_n(\tau)\} \leq B/\epsilon. \tag{A.8}$$

□

Let us now consider a single-hop network with time varying connectivity, which is studied in [6]. In this network, there is an AP serving N clients. Each client n generates one packet with probability a_n in each time slot, and packets not served by the AP are retained in its queue. We consider the ON/OFF channel model where each client n has an ON channel with the AP with probability p_n in each time slot. In each time slot, the AP can either transmit a packet for a client with an ON channel, or idle.

For this system, we can define $D_n(t)$ as the queue length of client n at time t. We first study whether it is possible to make the system strongly stable. A subset S of clients generates, on average, a total number of $\sum_{n \in S} a_n$ packets per time slot. In a time slot t, the AP can transmit a packet for a client in S only if at least one of the clients in S has an ON channel, which happens with probability $1 - \prod_{n \in S}(1 - p_n)$. Therefore, we can immediately obtain a necessary condition:

Lemma A.3 *There exists a policy that makes the system strongly stable only if, for all subsets S, $\sum_{n \in S} a_n < 1 - \prod_{n \in S}(1 - p_n)$.*

We now present a simple policy for this system. In this policy, the AP schedules the client with the longest queue from those that have ON channels. This policy is called the *longest connected queue* (LCQ) policy or the *maximum weight* policy. This policy is *throughput optimal* in the sense that it makes a system stable whenever possible.

Theorem A.4 *The LCQ policy makes the system stable if and only if, for all subsets S, $\sum_{n \in S} a_n < 1 - \prod_{n \in S}(1 - p_n)$.*

Proof. Using Theorem A.2, it suffices to show that there exists $B > 0, \epsilon > 0$ such that $\Delta(t) \le B - \epsilon \sum_n D_n(t)$, for all t. Without loss of generality, we assume that $D_1(t) \ge D_2(t) \ge \cdots \ge D_N(t)$. Using the LCQ policy, client n is scheduled if it has an ON channel, and all clients in $1-(n-1)$ have OFF channels, which happens with probability $p_n \prod_{i=1}^{n-1}(1 - p_i)$. Therefore, we have $E\{D_n(t + 1)|D(t)\} = D_n(t) + a_n - p_n \prod_{i=1}^{n-1}(1 - p_i)$. Let $\delta := \min_S(1 - \prod_{n \in S}(1 - p_n) - \sum_{n \in S} a_n)$. We now have

$$\Delta(t) = E\{\frac{1}{2}\sum_n D_n(t+1)^2 - \frac{1}{2}\sum_n D_n(t)^2 | D(t)\} \tag{A.9}$$

$$= E\{\frac{1}{2}\sum_n (D_n(t+1) + D_n(t))(D_n(t+1) - D_n(t))|D(t)\} \tag{A.10}$$

$$\le \sum_n D_n(t)(a_n - p_n \prod_{i=1}^{n-1}(1 - p_i)) + B \tag{A.11}$$

$$= \sum_{n=1}^{N}[(D_n(t) - D_{n+1}(t))\sum_{i=1}^{n}(a_i - p_i \prod_{j=1}^{i-1}(1 - p_j))] + B \tag{A.12}$$

$$= \sum_{n=1}^{N}\{(D_n(t) - D_{n+1}(t))(\sum_{i=1}^{n} a_i + \sum_{i=1}^{n}[\prod_{j=1}^{i}(1 - p_j) - \prod_{j=1}^{i-1}(1 - p_i)])\} + B \tag{A.13}$$

$$= \sum_{n=1}^{N}\{(D_n(t) - D_{n+1}(t))(\sum_{i=1}^{n} a_i + \sum_{i=1}^{n}[\prod_{j=1}^{i}(1 - p_j) - \prod_{j=1}^{i-1}(1 - p_j)])\} + B \tag{A.14}$$

$$= -\delta \sum_{n-1}^{N}\{(D_n(t) - D_{n+1}(t))\} + B \tag{A.15}$$

$$= -\delta D_1(t) + B \tag{A.16}$$

$$= -\frac{\delta}{N}\sum_{n=1}^{N} D_n(t) + B, \tag{A.17}$$

which completes the proof. □

APPENDIX B

Incentive Compatible Auction Design

Consider an item being auctioned with N bidders competing for it. Each bidder n has a secret valuation v_n for the item. In the auction, each bidder makes a bid b_n to the auctioneer, based on which the auctioneer sells the item to a bidder and charges the winning bidder some price. We use w to denote the winning bidder, and p_w to denote the amount of money that w is charged. The net utility of the winning bidder can then be written as $v_w - p_w$, while the net utilities of other bidders are zero.

Let us now discuss the mechanism used by the auctioneer, which includes choosing the winning bidder as well as the amount of money w is charged. The most intuitive and widely used mechanism is for the auctioneer to simply choose the winning bidder as one that offers the highest bid, and charge it the amount of its bid. That is, the auctioneer chooses $w := \arg\max\{b_n\}$ and $p_w := b_w$. Now, consider a special case where there are only two bidders with $v_1 = 10$ and $v_2 = 1$. Further, assume that bidder 2 bids $b_2 = v_2 = 1$. If bidder 1 bids its true valuation to the item, i.e., $b_1 = v_1 = 10$, it wins the auction and needs to pay $p_1 = b_1 = 10$. As a result, its net utility is $v_1 - p_1 = 0$. Instead, bidder 1 can still win the auction by bidding $b_1 = 1.01$, in which case it is charged $p_1 = 1.01$ and has a net utility of $10 - 1.01 = 8.99$.

From the above example, it is clear that the optimal bid that maximizes the net utility of a bidder depends on not only its own valuation, but also the secret valuations and bids of other bidders. For auctions that involve more items and complicated relations among the items, analyzing the optimal bidding strategy can become a huge overhead for the bidder, hence discouraging bidders from joining the auction. Further, the strategic behaviors of bidders makes it difficult for the auctioneer to project the outcome of each auction. Thus, it is more preferable to design an *incentive compatible auction mechanism* under which the net utility of each bidders n is maximized by bidding $b_n = v_n$, regardless of the bids of other bidders.

For auction involving only one item, such an incentive compatible auction is the well known *Vickrey's second price auction:*

Definition B.1 Vickrey's second price auction Let the winning bidder be the one that has the highest bid, with its assessed charge being the second highest bid. That is, $w = \arg\max\{b_n\}$ and $p_w = \max_{n \neq w} b_n$.

Theorem B.2 *Vickrey's second price auction is incentive compatible.*

Proof. Assume that bidder n wins by bidding $b_n = v_n$. We have $v_n \geq \max_{m \neq n} b_m$, and the net utility of n when it wins is $v_n - \max_{m \neq n} b_m \geq 0$, regardless of its actual bid. On the other hand, if n does not bid its true valuation and loses the auction, its net utility is zero, which is no larger than $v_n - \max_{m \neq n} b_m$. Hence, the net utility of n is maximized by bidding $b_n = v_n$.

Next consider the case when bidder n loses by bidding $b_n = v_n$, implying that $v_n \leq \max_{m \neq n} b_m$. If n chooses a bid that wins the auction, it needs to pay $\max_{m \neq n} b_m$, and its net utility is $v_n - \max_{m \neq n} b_m \leq 0$. Thus, in this case, the net utility of n is also maximized by bidding $b_n = v_n$. $\qquad\square$

In the Vickrey's second price auction, a bidder n wins if its bid is larger than $\max_{m \neq n} b_m$, which is the amount of charge it needs to pay when it wins. Define the *critical value* of client n as the smallest value it needs to bid in order to win the auction. Vickrey's second price auction then charges the winning bidder its critical value. This design can be extended to more complicated auctions.

Consider an auction where the auctioneer can select a subset of bidders among $S_1, S_2, \ldots$ as winning bidders. Chapter 6 describes several applications of this kind of auctions. The following *Vickrey-Clarke-Groves* (VCG) mechanism is incentive compatible for this auction:

Definition B.3   Given the bids from bidders, the VCG mechanism does the following:

1. It selects the subset S of bidders that maximizes $\sum_{n \in S} b_n$ as winning bidders.

2. For each bidder n in S, it charges n the critical value of n. In this scenario, the critical value of n is $\max_{S' \in \{S_1, S_2, \ldots\}, n \notin S'} \left(\sum_{m \in S'} b_m \right) - \sum_{m \in S, m \neq n} b_n$.

Theorem B.4   *The VCG mechanism is incentive compatible.*

The proof of this theorem is very similar to that of Theorem B.2, and is hence omitted.

Bibliography

[1] Cisco, "Cisco visual networking index: Global mobile data traffic forecast update, 2011–2016," 2012. 1

[2] K. Jain, J. Padhye, V. N. Padmanabhan, and L. Qiu, "Impact of interference on multi-hop wireless network performance," in *Proceedings of the 9th annual international conference on Mobile computing and networking*, MobiCom '03, 2003. 1

[3] T. Stockhammer, H. Jenkac, and G. Kuhn, "Streaming video over variable bit-rate wireless channels," *IEEE Trans. on Multimedia*, vol. 6, pp. 268–277, April 2004. DOI: 10.1109/TMM.2003.822795. 2

[4] S. H. Kang and A. Zakhor, "Packet scheduling algorithm for wireless video streaming," in *PV*, 2002. 2

[5] Q. Li and M. van der Schaar, "Providing adaptive QoS to layered video over wireless local area networks through real-time retry limit adaptation," *IEEE Trans. on Multimedia*, vol. 6, no. 2, pp. 278–290, 2004. DOI: 10.1109/TMM.2003.822792. 2

[6] L. Tassiulas and A. Ephremides, "Dynamic server allocation to parallel queues with randomly varying connectivity," *IEEE Trans. on Information Theory*, vol. 39, no. 2, pp. 89–103, 1993. DOI: 10.1109/18.212277. 2, 94

[7] M. Neely, "Delay analysis for max weight opportunistic scheduling in wireless systems," in *Proc. of Allerton Conf.*, 2008. DOI: 10.1109/TAC.2009.2026943. 2

[8] S. Shakkottai and A. L. Stolyar, "Scheduling algorithms for a mixture of real-time and non-real-time data in HDR," in *Proc. of 17th International Teletraffic Congress (ITC-17)*, 2001. DOI: 10.1016/S1388-3437(01)80170-0. 2

[9] K. B. Johnsson and D. C. Cox, "An adaptive cross-layer scheduler for improved QoS support of multiclass data services on wireless systems," *IEEE J. on Selected Areas in Communications*, vol. 23, no. 2, 2005. DOI: 10.1109/JSAC.2004.839381. 2

[10] A. Dua and N. Bambos, "Deadline constrained packet scheduling for wireless networks," in *62nd IEEE VTC Fall*, 2005. DOI: 10.1109/VETECF.2005.1557499. 2

[11] V. Raghunathan, V. Borkar, M. Cao, and P. Kumar, "Index policies for real-time multicast scheduling for wireless broadcast systems," in *Proc. of IEEE INFOCOM*, 2008. DOI: 10.1109/INFOCOM.2008.217. 2, 3

[12] S. Shakkottai and R. Srikant, "Scheduling real-time traffic with deadlines over a wireless channel," *Wireless Networks*, vol. 8, Jan. 2002. DOI: 10.1023/A:1012763307361. 2

[13] A. L. Stolyar and K. Ramanan, "Largest weighted delay first scheduling: Large deviations and optimality," *Ann. Appl. Probab.*, vol. 11, no. 1, 2001. DOI: 10.1214/aoap/998926986. 2

[14] T. Kawata, S. Shin, A. G. Forte, and H. Schulzrinne, "Using dynamic PCF to improve the capacity for VoIP traffic in IEEE 802.11 networks," in *Proc. of IEEE WCNC*, 2005. DOI: 10.1109/WCNC.2005.1424751. 2

[15] H. Fattah and C. Leung, "An overview of scheduling algorithms in wireless multimedia networks," *IEEE Wireless Communications*, vol. 9, pp. 76–83, Oct. 2002. DOI: 10.1109/MWC.2002.1043857. 2

[16] Y. Cao and V. Li, "Scheduling algorithms in broadband wireless networks," *Proceedings of the IEEE*, vol. 89, pp. 76–87, Jan. 2001. DOI: 10.1109/5.904507. 2

[17] Y. Xiao and H. Li, "Evaluation of distributed admission control for the IEEE 802.11e EDCA," *Communications Magazine, IEEE*, vol. 42, Sept. 2004. DOI: 10.1109/MCOM.2004.1336720. 2

[18] D. Pong and T. Moors, "Call admission control for IEEE 802.11 contention access mechanism," in *Proc. of GLOBECOM*, 2003. DOI: 10.1109/GLOCOM.2003.1258225. 2

[19] S. Garg and M. Kappes, "Admission control for VoIP traffic in IEEE 802.11 networks," in *Proc. of GLOBECOM*, 2003. DOI: 10.1109/GLOCOM.2003.1258888. 2

[20] H. Zhai, X. Chen, and Y. Fang, "A call admission and rate control scheme for multimedia support over IEEE 802.11 wireless LANs," *Wireless Networks*, vol. 12, August 2006. DOI: 10.1007/s11276-006-6545-y. 2

[21] S. Shin and H. Schulzrinne, "Call admission control in IEEE 802.11 WLANs using QP-CAT," in *Proc. of INFOCOM*, 2008. DOI: 10.1109/INFOCOM.2008.123. 2

[22] D. Gao, J. Cai, and K. Ngan, "Admission control in IEEE 802.11e wireless LANs," *IEEE Network*, pp. 6–13, July/August 2005. DOI: 10.1109/MNET.2005.1470677. 2

[23] D. Niyato and E. Hossain, "Call admission control for QoS provisioning in 4G wireless networks: issues and approaches," *IEEE Network*, pp. 5–11, September/October 2005. DOI: 10.1109/MNET.2005.1509946. 2

[24] M. Ahmed, "Call admission control inwireless networks: A comprehensive survey," *IEEE Communications Surveys*, vol. 7, no. 1, pp. 50–69, 2005. DOI: 10.1109/COMST.2005.1423334. 2

[25] F. Kelly, "Charging and rate control for elastic traffic," *European Trans. on Telecommunications*, vol. 8, pp. 33–37, 1997. DOI: 10.1002/ett.4460080106. 3, 41

[26] F. Kelly, A. Maulloo, and D. Tan, "Rate control in communication networks: shadow prices, proportional fairness and stability," *Journal of the Operational Research Society*, vol. 49, pp. 237–252, 1998. DOI: 10.2307/3010473. 3

[27] X. Lin and N. Shroff, "Utility maximization for communication networks with multi-path routing," *IEEE Trans. on Automated Control*, vol. 51, no. 5, pp. 766–781, 2006. DOI: 10.1109/TAC.2006.875032. 3

[28] M. Xiao, N. Shroff, and E. Chong, "A utility-based power-control scheme in wireless cellular systems," *IEEE/ACM Trans. on Networking*, vol. 11, no. 2, pp. 210–221, 2003. DOI: 10.1109/TNET.2003.810314. 3

[29] Y. Cao and V. Li, "Utility-oriented adaptive QoS and bandwidth allocation in wireless networks," in *Proc. of ICC*, 2002. DOI: 10.1109/ICC.2002.997403. 3

[30] G. Bianchi, A. Campbell, and R. Liao, "On utility-fair adaptive services in wireless networks," in *Proc. of IWQoS*, pp. 256–267, 1998. DOI: 10.1109/IWQOS.1998.675246. 3

[31] X. Zhang and Q. Du, "Cross-layer modeling for QoS-driven multimedia multicast/broadcast over fading channels in mobile wireless networks," *IEEE Communications Magazine*, vol. 45, pp. 62–70, Aug. 2007. DOI: 10.1109/MCOM.2007.4290316. 3

[32] P. Gopala and H. E. Gamal, "On the throughput-delay tradeoff in cellular multicast," in *Proceedings of the Symposium on Information Theory in WirelessCom*, 2005. 3

[33] S. Zhou and L. Ying, "On delay constrained multicast capacity of large-scale mobile ad-hoc networks," in *Proc. of IEEE INFOCOM*, 2010. DOI: 10.1109/INFCOM.2010.5462257. 3

[34] P. Chaporkar and A. Proutiere, "Adaptive network coding and scheduling for maximizing throughput in wireless networks," in *Proceedings of ACM MobiCom*, pp. 135–146, 2007. DOI: 10.1145/1287853.1287870. 3

[35] M. Ghaderi, D. Towsley, and J. Kurose, "Reliability gain of network coding in lossy wireless networks," in *Proc. of IEEE INFOCOM*, pp. 2171–2179, 2008. DOI: 10.1109/INFOCOM.2008.284. 3

[36] D. Nguyen, T. Tran, T. Nguyen, and B. Bose, "Wireless broadcast using network coding," *IEEE Transactions on Vehicular Technology*, vol. 58, pp. 914–925, Feb. 2009. DOI: 10.1109/TVT.2008.927729. 3

[37] D. Lucani, M. Medard, and M. Stojanovic, "Systematic network coding for time-division duplexing," in *Proceedings of IEEE ISIT*, pp. 2403–2407, 2010. DOI: 10.1109/ISIT.2010.5513768. 3

[38] U. Kozat, "On the throughput capacity of opportunistic multicasting with erasure codes," in *Proceedings of IEEE INFOCOM*, pp. 520–528, 2008. DOI: 10.1109/INFOCOM.2008.100. 3

[39] W.-L. Yeow, A. T. Hoang, and C.-K. Tham, "Minimizing delay for multicast-streaming in wireless networks with network coding," in *Proceedings of IEEE INFOCOM*, pp. 190–198, 2009. DOI: 10.1109/INFCOM.2009.5061921. 3

[40] A. Eryilmaz, A. Ozdaglar, and M. Medard, "On delay performance gains from network coding," in *Proc. of CISS*, pp. 864–870, 2006. 3

[41] L. Ying, S. Yang, and R. Srikant, "Coding achieves the optimal delay-throughput trade-off in mobile ad-hoc networks: Two-dimensional i.i.d. mobility model with fast mobiles," in *Proc. of WiOpt*, pp. 1–10, 2007. DOI: 10.1109/WIOPT.2007.4480024. 3

[42] X. Li, C.-C. Wang, and X. Lin, "Throughput and delay analysis on uncoded and coded wireless broadcast with hard deadline constraints," in *Proc. of IEEE INFOCOM*, 2010. DOI: 10.1109/INFCOM.2010.5462258. 3

[43] W. Pu, C. Luo, F. Wu, and C. W. Chen, "QoS-driven network coded wireless multi-cast," *IEEE Transactions on Wireless Communications*, vol. 8, pp. 5662–5670, Nov. 2009. DOI: 10.1109/TWC.2009.090203. 3

[44] H. Gangammanavar and A. Eryilmaz, "Dynamic coding and rate-control for serving deadline-constrained traffic over fading channels," in *Proc. of IEEE ISIT*, pp. 1788–1792, 2010. DOI: 10.1109/ISIT.2010.5513290. 3

[45] J. Liu, *Real-Time Systems*. Prentice Hall, 2000. 4

[46] C. L. Liu and J. W. Layland, "Scheduling algorithms for multiprogramming in a hard-real-time environment," *J. ACM*, vol. 20, no. 1, 1973. DOI: 10.1145/321738.321743. 4

[47] J.-T. Chung, J. W. S. Liu, and K.-J. Lin, "Scheduling periodic jobs that allow impre-cise results," *IEEE Transactions on Computers*, vol. 39, pp. 1156–1174, September 1990. DOI: 10.1109/12.57057. 4

[48] J. W. Liu, K.-J. Lin, W.-K. Shih, and A. C.-S. Yu, "Algorithms for scheduling imprecise computations," *Computers*, vol. 24, pp. 58–68, May 1991. 4

[49] W.-K. Shih and J. W. Liu, "Algorithms for scheduling imprecise computations with timing constraints to minimize maximum error," *IEEE Transactions on Computers*, vol. 44, pp. 466–471, March 1995. DOI: 10.1109/12.372040. 4

[50] P. H. Feiler and J. J. Walker, "Adaptive feedback scheduling of incremental and design-to-time tasks," in *Proceedings of the 23rd International Conference on Software Engineering*, pp. 318–326, 2001. DOI: 10.1109/ICSE.2001.919105. 4

[51] P. Mejia-Alvarez, R. Melhem, and D. Mosse, "An incremental approach to scheduling during overloads in real-time systems," in *Proceedings of the 21st IEEE Real-Time Systems Symposium*, pp. 283–293, 2000. DOI: 10.1109/REAL.2000.896017. 4

[52] J.-M. Chen, W.-C. Lu, W.-K. Shih, and M.-C. Tang, "Imprecise computations with deferred optional tasks," *Journal of Information Science and Enginnering*, vol. 25, no. 1, pp. 185–200, 2009. 4

[53] M. Zu and A. M. K. Chang, "Real-time scheduling of hierarchical reward-based tasks," in *Proceedings of the 9th IEEE Real-Time and Embedded Technology and Applications Symposium*, pp. 2–9, 2003. DOI: 10.1109/RTTAS.2003.1203031. 4

[54] H. Aydin, R. Melhem, D. Mosse, and P. Mejia-Alvarez, "Optimal reward-based scheduling for periodic real-time tasks," *IEEE Transactions on Computers*, vol. 50, pp. 111–130, February 2001. DOI: 10.1109/12.908988. 4

[55] M. Amirijoo, J. Hansson, and S. H. Son, "Specification and management of QoS in real-time databases supporting imprecise computations," *IEEE Transactions on Computers*, vol. 55, pp. 304–319, March 2006. DOI: 10.1109/TC.2006.45. 4

[56] J. K. Dey, J. Kurose, and D. Towsley, "On-line scheduling policies for a class of IRIS (increasing reward with increasing service) real-time tasks," *IEEE Transactions on Computers*, vol. 45, pp. 802–813, July 1996. DOI: 10.1109/12.508319. 4

[57] H. Cam, "An on-line scheduling policy for IRIS real-time composite tasks," *The Journal of Systems and Software*, vol. 52, pp. 25–32, 2000. DOI: 10.1016/S0164-1212(99)00130-2. 4

[58] I.-H. Hou, V. Borkar, and P. R. Kumar, "A theory of QoS for wireless," in *Proc. of IEEE INFOCOM*, 2009. DOI: 10.1109/INFCOM.2009.5061954. 5

[59] I.-H. Hou and P. R. Kumar, "Admission control and scheduling for QoS guarantees for variable-bit-rate applications on wireless channels," in *Proc. of ACM MobiHoc*, pp. 175–184, 2009. DOI: 10.1145/1530748.1530772. 5

[60] I.-H. Hou and P. R. Kumar, "Scheduling heterogeneous traffic over fading wireless channels," in *Proc. of IEEE INFOCOM*, 2010. DOI: 10.1109/INFCOM.2010.5462090. 5

[61] I.-H. Hou and P. R. Kumar, "Utility maximization for delay constrained qos in wireless," in *Proc. of IEEE INFOCOM*, 2010. DOI: 10.1109/INFCOM.2010.5462070. 5

[62] I.-H. Hou and P. R. Kumar, "Utility-optimal scheduling in time-varying wireless networks with delay constraints," in *Proc. of ACM MobiHoc*, 2010. DOI: 10.1145/1860093.1860099. 5

[63] J. Jaramillo and R. Srikant, "Optimal scheduling for fair resource allocation in ad hoc networks with elastic and inelastic traffic," *Networking, IEEE/ACM Transactions on*, vol. 19, pp. 1125–1136, aug. 2011. DOI: 10.1109/TNET.2010.2100083. 5, 71

[64] I.-H. Hou and P. R. Kumar, "Broadcasting delay-constrained traffic over unreliable wireless links with network coding," in *Proc. of ACM MobiHoc*, 2011. DOI: 10.1145/2107502.2107508. 6

[65] M. Csorgo, "On the strong law of large numbers and the central limit theorem for martingales," *Trans. Amer. Math. Soc.*, vol. 131, pp. 259–275, 1968. DOI: 10.2307/1994694. 11

[66] S. Resnick, *A Probability Path*. Birkhauser Boston, 1998. 11

[67] D. Blackwell, "An analog of the minimax theorem for vector payoffs," *Pacific J. Math*, vol. 6, no. 1, 1956. DOI: 10.2140/pjm.1956.6.1. 17, 48

[68] M. Bazaraa, H. Sherali, and C. Shetty, *Nonlinear programming: theory and algorithms*. Wiley-Interscience, 2006. DOI: 10.1002/0471787779. 42, 62

[69] X. Lin and N. Shroff, "Joint rate control and scheduling in multihop wireless networks," in *Proc. of IEEE CDC*, pp. 1484–1489, 2004. DOI: 10.1109/CDC.2004.1430253. 59

[70] Y. Chow and H.Teicher, *Probability theory, independence, interchangeability, martingales*. Springer-Verlag, 1988. 61

[71] J. Neveu, *Discrete parameter maringales*. North-Holland, 1975. 62

[72] W. Vickrey, "Counterspeculation, auctions and competitive sealed tenders," *J. of Finance*, vol. 16, pp. 8–37, 1961. DOI: 10.1111/j.1540-6261.1961.tb02789.x. 63

[73] E. Clarke, "Multipart pricing of public goods," *Public Choice*, vol. 11, pp. 17–33, Sep. 1971. DOI: 10.1007/BF01726210. 63

[74] T. Groves, "Incentives in teams," *Econometrica*, vol. 41, pp. 617–631, Jul. 1973. DOI: 10.2307/1914085. 63

[75] N. Nisan, T. Roughgarden, E. Tardos, and V. Vazirani, *Algorithmic game theory*. Cambridge University Press, 2007. DOI: 10.1017/CBO9780511800481. 64

[76] T. Cormen, C. Leiserson, R. Rivest, and C. Stein, *Introductions to Algorithms (2ed)*. The MIT Press, 2001. 81

[77] Texas Instruments, "Low power advantage of 802.11a/g vs. 802.11b," Dec. 2003. 87

Authors' Biographies

I-HONG HOU

I-Hong Hou received his B.S. in Electrical Engineering from National Taiwan University in 2004, and his M.S. and Ph.D. in Computer Science from University of Illinois, Urbana-Champaign in 2008 and 2011, respectively.

In 2012, he joined the Department of Electrical and Computer Engineering at Texas A&M University, where he is currently an Assistant Professor. His research interests include wireless networks, wireless sensor networks, real-time systems, distributed systems, and vehicular ad hoc networks.

Dr. Hou received the C.W. Gear Outstanding Graduate Student Award from the University of Illinois at Urbana-Champaign, and the Silver Prize in the Asian Pacific Mathematics Olympiad.

P. R. KUMAR

P. R. Kumar obtained his B. Tech. degree in Electrical Engineering (Electronics) from I.I.T. Madras in 1973, and his M.S. and D.Sc. degrees in Systems Science and Mathematics from Washington University, St. Louis, in 1975 and 1977, respectively. From 1977-1984 he was a faculty member in the Department of Mathematics at the University of Maryland Baltimore County. From 1985-2011 he was a faculty member in the Department of Electrical and Computer Engineering and the Coordinated Science Laboratory at the University of Illinois. Currently, he is at Texas A&M University, where he holds the College of Engineering Chair in Computer Engineering.

Kumar has worked on problems in game theory, adaptive control, stochastic systems, simulated annealing, neural networks, machine learning, queueing networks, manufacturing systems, scheduling, wafer fabrication plants, and information theory. His research is currently focused on power systems, wireless networks, and cyberphysical systems.

Kumar is a member of the National Academy of Engineering of the USA, and the Academy of Sciences of the Developing World. He was awarded an honorary doctorate by ETH, Zurich. He received the IEEE Field Award for Control Systems, the Donald P. Eckman Award of the American Automatic Control Council, the Fred W. Ellersick Prize of the IEEE Communications Society, and the Outstanding Contribution Award of ACM SIGMOBILE. He is a Fellow of IEEE. He was a Guest Chair Professor and Leader of the Guest Chair Professor Group on Wireless Communication and Networking at Tsinghua University, Beijing, China. He is a D. J. Gandhi Distinguished Visiting Professor at IIT Bombay. He is an Honorary Professor at IIT Hyderabad. He was awarded the Distinguished Alumnus Award from IIT Madras, the Alumni Achievement

Award from Washington University in St. Louis, and the Daniel C. Drucker Eminent Faculty Award from the College of Engineering at the University of Illinois.